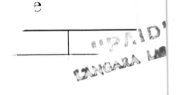

Structured Analysis

PRENTICE-HALL SOFTWARE SERIES
Brian W. Kernighan, advisor

Structured Analysis

by Victor Weinberg

Preface by Edward Yourdon

Prentice-Hall, Inc., Englewood Cliffs, New Jersey 07632

Library of Congress Cataloging in Publication Data

Weinberg, Victor (date)
   Structured analysis.

   Includes bibliographical references and index.
      1. Electronic digital computers—Programming.
   2. System analysis.   I. Title.
   QA76.6.W453 1980   003   79-23508
   ISBN 0-13-854414-X

Cover Design by Suzanne Behnke
Manufacturing Buyer: Gordon Osbourne

Printed in the United States of America

10 9 8 7 6 5 4 3

PRENTICE-HALL INTERNATIONAL, INC., *London*
PRENTICE-HALL OF AUSTRALIA PTY. LIMITED, *Sydney*
PRENTICE-HALL OF CANADA, LTD., *Toronto*
PRENTICE-HALL OF INDIA PRIVATE LIMITED, *New Delhi*
PRENTICE-HALL OF JAPAN, INC., *Tokyo*
PRENTICE-HALL OF SOUTHEAST ASIA PTE. LTD., *Singapore*
WHITEHALL BOOKS LIMITED, *Wellington, New Zealand*

In memory of my father

and

to my mother

Acknowledgments

I acknowledge my co-workers, friends, and the many attendees of the seminars I have taught for their ideas, criticism, technical assistance, and overall support. I identify no names only to omit no names. I further acknowledge the following philosophical concept from Tibetan mysticism, which served as a source of strength during moments of frustration and occasional despair:

It is in the unseen providence of things that sometimes our greatest difficulties are our best opportunities.

VICTOR WEINBERG

Preface

It is a great pleasure for me to introduce Victor Weinberg's book on structured analysis. It provides, I think, a much-needed addition to the meager library of books on systems analysis.

During the past three years, a whole new discipline of systems analysis has emerged — largely through the work of people in our company, and in such organizations as SofTech, TRW, and the ISDOS project at the University of Michigan. Victor Weinberg has been involved with this new discipline from its inception, and has made valuable contributions as a teacher, as a consultant, and as a researcher.

For me, the best thing about Victor's book is its marriage of new analysis techniques and conventional analysis techniques. With all of the current interest in data flow diagrams, data dictionaries, HIPO, and other tools of structured analysis, it is easy to forget the importance of such classical problem areas as the definition of user objectives, and the establishment of a formal consensus between the user, the analyst, and the development team: The last five chapters of the book concentrate on these problems of analysis. In short, the book presents, in detail, a formal project methodology that incorporates all of the structured documentation techniques, but still retains the essence of good, classical project management and control.

All who read *Structured Analysis* will be struck by the same thought that came to me: These are the words of a man who has actually *done* it. I often suspect, when reading a book on programming or systems analysis, that the author is proposing some interesting theories, but that he or she has never really struggled through a *real* project, and dealt with *real* users. Having worked with Victor for the past three years, and having watched him in action on project after project, I already know that he knows what he's talking about. But it's a pleasure to see the same real-world "savvy" in his book, too: Example after example and anecdote after anecdote are used to illustrate the principles and the philosophies he espouses.

Systems analysts, in general, seem to be a skeptical and slightly cynical lot — those who survive have a lot of scars, and they're the first to tell you that they've seen it all, and they've done it all. Yet even the most battle-scarred veteran will find *Structured Analysis* a book that he can enjoy, and that he can learn from.

Happy reading!

New York
April 1978

Edward Yourdon
President, YOURDON inc.

Structured Analysis

Introduction

"Structured analysis" provides both useful insights into the problems of systems analysis and graphic tools to minimize some of the inherent communication problems. Yet, some people feel that structured analysis is little more than common sense, perhaps because they believe that the subject of systems analysis defies structuring. Or, perhaps they feel that effective guidelines, techniques, and strategies needed to stand up to the enormous variety of problems, personalities, and environments that systems analysts face cannot be created. Perhaps analysts, who almost by definition are battle-worn, experienced data processing professionals, do not like to think that they have not been doing a good job.

Experienced users, analysts, and managers do not need to be convinced that current analytical methods are primitive. Their experiences with projects — which have been late, over-budget, inflexible to change, and unacceptable to the users — support the point better than anything that could be said. What is less clear is whether structured analysis or any other new techniques include the insights, techniques, and strategies to make systems analysts more effective in all aspects of their systems development efforts.

Perhaps the most critical problem in systems analysis is that, *often, neither the user nor the systems analyst knows what is required or desirable.* To solve this problem, users and systems analysts should work together, but too often they view each other as adversaries. They may work for different departments and have different ambitions and perspectives. Yet, it should be remembered that they are all working toward the same goal, a successful systems development effort. The analyst's objective in presenting solutions to users should be to get the users to

1

understand and believe what they do not understand and may not want to believe. Of course, *telling someone something is not equivalent to his understanding and believing it.* "You have not converted a man because you have silenced him."* Yet, that understanding and belief are critical to the success of the systems analyst's efforts.

I am reminded of the story of two analysts who worked independently on the same problem. They used the same approaches and came up with the same valid conclusions, differing only in the way they presented their conclusions to the users. One analyst held a meeting with all of the users, walking them through the analytic process leading to the conclusion. Users periodically interrupted the presentation to get clarifications on the technical subject matter. The resulting dialogue cleared up all misconceptions. After the conclusions were presented, the users enthusiastically agreed to implement all of the systems analyst's suggestions.

The second analyst distributed a thick, technical document to all of the users. At the end of the document, the analyst noted that as a result of his findings, certain departments would be required to change their operational procedures . . . *immediately.* Within a week, the second analyst received a number of memos from the users, all with essentially the same response: They did not understand the document, could not have confidence in the analyst's findings, and refused to implement the new procedures as presented.

One conclusion to be drawn from this story is that the *way* an idea is presented may be as important as the idea itself in terms of its ultimate acceptance. What we need is to get our audience involved in not just the solution, but also in the thought process leading to the solution. The audience is much more likely to accept the validity of the conclusion if they understand it, rather than simply being told it . . . even if they are told by some "authority."

---

*John Viscount Morley, *On Compromise* (London: Chapman and Hall, 1874).

The point is that the systems analyst's job involves considerably more than technical expertise. The analyst, in order to be effective, must be oriented to people. Not only must the analyst create solutions, but he also must communicate these solutions effectively. He must create an understanding, a rapport, and a feeling of collaboration with the entire user community.

Let's take a look at another story, this time outside the computer field. I recently reread an article about Marco Aldaco, an architect who produced spectacular results without using standard architectural methods. He said:

> The error of today's architects is that
>   they work in offices with T-squares.
> They might as well be working in factories.
> They plan houses as if they were making
>   the same Ford car over and over . . . .
>
> I cannot build a preconceived house . . .
> I must understand the land . . .
> feel the cold at night . . .
> feel how the winds blow . . .
> see how the sun moves and the birds fly . . .
> think about the history of the place and of the people . . . .
>
> The personality of the owner is very important . . . .
> I need someone who can help me — through
>   his personality — build a house.
> The owner must consider architecture a
>   work of art.
> He is my collaborator.*

I find these ideas to be extremely relevant to our systems development efforts. While some might say, "It's just common sense," we tend to forget why we do the things we do. Perhaps more importantly, we lose track of what to emphasize among all our activities. An important theme of this book is that the analyst must "go to the land" to get really in touch with the users, their problems, and their objectives. Working in some

---

*Allen Carter, "Primitive Sophistication on the Costa de Careyes," *Architectural Digest,* Vol. 33, No. 1 (July/August 1976), pp. 45-48. Reprinted from *Architectural Digest.* Copyright © John C. Brasfield Publishing Corp., 1976.

back room using memos and the telephone, the analyst will get only a secondhand view of the problem.

The back-room approach is not a substitute for firsthand data gathering. The analyst must touch, feel, and have extensive, personal contact with the environment. By doing so, the analyst will be able to obtain sufficient information to test the validity of preconceived notions. Furthermore, he will open the lines of communication with the user community. *The most successful projects are truly joint efforts in which both analysts and users make significant contributions:* Users bring to the problem an indispensable business perspective; analysts bring objectivity, technical expertise, and analytical skills. They need each other's perspective and cooperation to do the best possible job.

A point that emerges here relates to the artistic nature of the systems analyst's job. We deal with people — their weaknesses and strengths, their needs and objectives. No checklist of procedures totally can replace the sensitivity, thoughtfulness, and imagination required in systems analysis. Weaknesses in these analytic requirements have been at the root of many of our failures in systems development. *Structured analysis attempts to identify these requirements and, whenever possible, to transform as much of the art into a craft that can be learned and applied.* As systems grow in complexity, analysts increasingly will be hardpressed to apply a disciplined approach utilizing effective tools. We must realize, as systems come and go, that we all are not Michelangelo.

My purpose in the pages that follow is to present a structured approach to systems development and to identify and clarify the role of the systems analyst. The book is intended primarily for systems analysts who have had little formal training in structured systems development, but it is particularly relevant as well to designers, project leaders, and users whose concern is the development, evaluation, and coordination of systems. In addition, programmers responsible for some analysis and design, and developers of systems standards and methodologies should find the book useful in those areas of their job that relate to structured systems development.

PART 1

Overview of Structured Analysis

1 Systems Analysis and the Systems Analyst

The users are interested in the whole system, not in just a part of the developmental process.

1.1 The state of the art

As an instructor of structured analysis, I normally begin a seminar by introducing myself to the class. Each attendee then is asked to identify his title, job responsibilities, and expectations of the seminar. While most attendees have the title of systems analyst, their responsibilities vary greatly, as do their training objectives. Some systems analysts come from user departments. Their job is to identify and represent the interests of their users. Some analysts are computer professionals, responsible for defining problems and designing solutions. Others define problems and develop functional specifications, but are not directly involved with the physical design of systems. Some are programmer/analysts who not only define problems and design solutions, but also implement the solutions. Others have responsibilities that are unrelated to the design and implementation of a particular computer system: They may be interested in developing standards and systems development methodologies, for example, or they may be concerned with developing improved project control strategies or techniques for cost/benefit analysis. Some are members of task forces, responsible for optimization and stabilization of a data processing environment.

Perhaps the one thing that these systems analysts have in common is that they have had almost no formal training in *how* to do analysis. Whenever I ask how they became systems analysts, a typical response is as follows:

> One day, my manager called me in for my yearly review. He said, "We like your work. You're getting a $1,500 raise. By the way, you're no longer a senior programmer. You're a systems analyst."

One seminar attendee, a professor from a midwestern university, commented that while universities have courses on how to program and design systems, most do not have any training program for systems analysis. Programming is an easily defined domain of work, with clear-cut rules and limitations. Students can be taught how to program just as they can be taught other crafts. The design of systems involves more imagination and demands a broader perspective of the environment than does the detailed work of programming. We all agree, however, that system designs should be flexible, efficient, and should provide the functional and operational capabilities defined in the system's specifications. For this reason, some universities now are reasonably successful in teaching students to design *maintainable* systems from existing functional specifications.

The subject of systems analysis, however, has been particularly difficult to teach because of its unique characteristics. Programming presupposes the existence of program specifications as the input to the process. Design requires functional specifications, i.e., the definition of objectives, requirements, and priorities of users. Systems analysis, by contrast, very often starts with almost no definition of the problem or of the environment where the problem exists. While educational institutions can create mock specifications to drive the design and programming processes, they are hard-pressed to simulate the politics and conflicting objectives and priorities of a real-world user environment. It is not sufficient to tell a class of systems analysts that Sam will be considered the user. *One of the basic problems of systems analysis is that the user often is a multiplicity of users, each with his own perspectives and ambitions.*

Another problem in teaching systems analysis is that while the boundaries of design and programming are relatively clear, the scope of systems analysis and the role of the systems analyst are subject to continual debate. In 1976, I attended a conference devoted to the skill requirements, tools, training, roles, and problems of the systems analyst. Approximately twenty papers were presented to an audience of about three hundred systems analysts. The titles of two papers particularly struck home because they seemed to indicate the state of the art of systems analysis: "Is Anyone Here a Systems Analyst?"[1] and "There's No Such Thing as a Systems Analyst."[2] Both papers suggested that the responsibilities of the systems analyst are so all-encompassing and ill-defined that no one person can possibly do the entire job. How, then, can anybody teach a subject that seems to defy definition?

I recall working as a consultant with a manager who frequently found herself in situations in which communication deteriorated due to a lack of clear definition of the basic terminology. One of her favorite comments was, "This is ridiculous. You're talking apples and I'm talking oranges." Rather than have this book float aimlessly in a sea of ill-defined terminology, let's get to the root of the problem by defining the terms associated with the problem before we attempt to determine solutions.

1.2 What a system is

Many people who work in the data processing field have difficulty understanding the word "system"; they refuse to accept it without a qualifying adjective.[3] For instance, they might argue that the statement, The system is up and running, is ambiguous. By itself, such a phrase surely is ambiguous; but we don't say such things in a vacuum. Rather, we speak in the context of a particular situation. If the context is sufficiently established, then we should understand the meaning of the words. For example, if we ask a telephone worker who is installing a telephone network at our office, "When will we have the system?" it is clear we are talking about a specific telephone system. If someone says, "You sound great" to a friend who is coughing and sneezing, and if the friend answers, "I've got a bug in my system," it is clear he or she is not talking about a technical problem in a computer system.

The word system in association with "analysis" and "analyst" refers to interdependent devices, people, rules, and/or procedures organized to form an integral whole to achieve a common purpose. Therefore, *to analyze a system is to identify the complex, its components (people, machines, rules, and procedures), and their interrelationships to determine objectives, requirements, priorities, and the extent to which they all have been satisfied.*

Certainly, the word system is abstract. We can photograph three apples to get a good picture of what apples are. We cannot, however, photograph three different systems to get a clear picture of what systems are. We can touch, feel, smell, and even eat apples. We cannot do any of these things with systems. Although systems come in all sizes and shapes and address sets of unique objectives, we should not be overly concerned with the abstract nature of the concept of a system. A more real problem with the terminology relates to the words analyst and analysis.

1.3 What analysis is

Does an analyst do analysis? This question is not as trivial as it seems. If analysis is defined as the process of examining, identifying, or separating a complex to define the relationship of its components, then the answer to the question is yes. A forms analyst, for example, identifies each individual component making up the complex and designs forms to satisfy all requirements. A methods analyst identifies who is involved in a project, how they relate to each other, and what their objectives are, and then sets up procedures by which these people will meet the objectives. A systems analyst may be required to examine an operational system to determine efficiency or the extent to which objectives are being met (both involve an examination of the system's complex and its components). In the examples presented, the analyst does analysis as long as analysis is defined broadly as a process of examination.

But in today's systems development environment, the word analysis also has a more specific meaning. Analysis frequently is used to describe the front-end phase of the systems development life cycle prior to the design phase. In this phase, problems and objectives are defined, tentative solutions proposed, and costs and benefits evaluated. (See Fig. 1.1.)

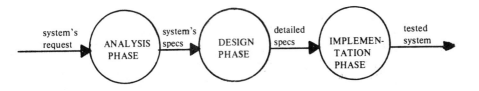

Figure 1.1. General phases of systems development.

If we define analysis as a front-end phase instead of a broad process of examination,[4,5] it is not so clear that an analyst does analysis. In the examples already described, the forms analyst, the methods analyst, and the systems analyst, doing post-evaluation work, are all involved in systems development phases other than front-end analysis. So we can see that whether an analyst does analysis depends on the specific nature of the analyst's job and how we define the word analysis.

One way out of the dilemma is to rename the front-end phase of the systems development life cycle. Alternatively, we could come up with a new buzzword to describe what an analyst does. Examination of problems, however, is the basic activity of analysts no matter what the specifics of the problems are, computer-related or not. While we should recognize that the existing terminology is misleading, the fact is it cannot be changed easily. Rather than reprogram ourselves to ignore the everyday meaning of analysis, we should clarify its relationship to the systems development life cycle. For example, we might distinguish the analysis phase (the front-end phase of the systems life cycle) from the analysis process (the examination of problems associated with the entire systems development effort). In doing so, it would be clear that the analysis process is included in, but is not limited to, the analysis phase.

1.4 Systems analysis and the systems analyst

Because we have such difficulty in defining the word analysis, it has been problematical for people in the data processing industry to define precisely the *function* of systems analysis and the *role* of the systems analyst. If we use the broad definition of analysis — i.e., the analysis process — then we can define the function of systems analysis as follows: *Systems analysis is the examination, identification, and evaluation of the components and interrelationships involved in systems from problem definition through maintenance and modification phases.* A simplified model of the systems analyst's role, performing these functions, is shown in Fig. 1.2.

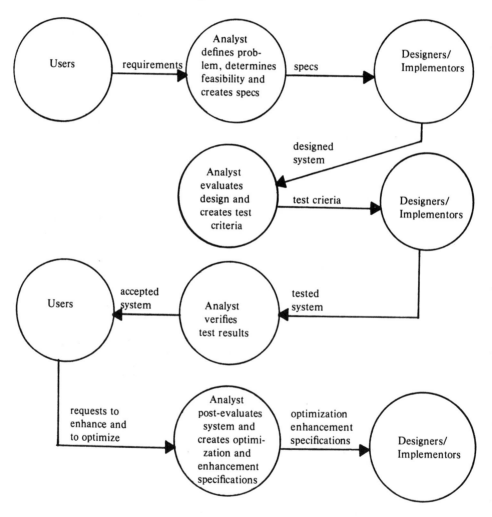

Figure 1.2. Analysis process showing systems analyst's role.

If instead we use the narrow definition of analysis, — i.e., the analysis phase — we can define *systems analysis as the examination of problems, objectives, requirements, priorities, and constraints in an environment, plus identification of cost estimates, benefits, and time requirements for tentative solutions.* The systems analyst's role in performing these functions is shown in Fig. 1.3. Both of these models (Figs. 1.2 and 1.3) show that the systems analyst interfaces with the system's *users* in order to define their requirements.

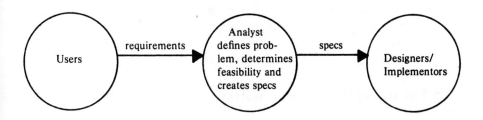

Figure 1.3. Analysis phase showing systems analyst's role.

To identify what systems analysis is, we should ask the following questions:

- What do users really want and need?

- Do they want and need their requirements to be defined properly, or do they want a system that meets these requirements?

- If they want a system that meets these requirements, who will act on their behalf to assure that these objectives and requirements are met?

- Assuming that users may not have sufficient technical expertise to evaluate thoroughly the systems produced, who will perform this function, if not the systems analyst?

These questions suggest that *the users are interested in the whole system, not in just a part of the developmental process.* The history of unsuccessful systems development projects has indicated that precise definitions of the problems and of the objectives of the users do not necessarily assure good systems. If anything, the history of systems development shows the weakness of systems analysts' having only partial responsibility. Part of a systems development process may be totally correct; but when all of the parts do not fit together properly, the system *as a whole* is a failure. What the users want is a successful system, and what they need is someone to take a broad approach to systems problem-solving that focuses on whole systems, not just on the individual parts of a process.[6]

1.5 The role of the systems analyst

The systems analyst traditionally has been thought of as interfacing with users and designers during the analysis phase, that is, only at the front end of the systems development life cycle. *What users need, however, is continuous evaluation of the systems development process to assure that design and implementation do not stray from the users' objectives, defined in earlier phases.* Regardless of traditional definitions of roles, systems analysts have been called upon to fill this need. In reality, their responsibilities extend throughout the entire systems development effort. While this role varies from project to project and from organization to organization, Fig. 1.4 identifies a typical breakdown of where the resources of the systems analyst are consumed in the systems development life cycle.[7]

Analysis	24%
Detailed design	26%
Programming/testing	30%
Training/implementation	13%
Post-evaluation	7%

Figure 1.4. Typical utilization of systems analyst's resources.

In the analysis phase, for example, systems analysts should begin by identifying the problem and the feasibility of solving it. Suggested activities follow:

1. Determine who the users are.
2. Determine how to gather information.
3. Collect and analyze the information.
4. State the problem.
5. Describe the current system.
6. Quantify the costs associated with the current system.
7. Summarize the current system's limitations and constraints.
8. Establish all preliminary objectives and requirements.
9. Formulate general alternative solutions.
10. Do preliminary cost/benefit analysis for each tentative physical solution.
11. Estimate the time and costs involved to develop alternative detailed solutions.
12. Summarize findings and state whether the problem warrants further developmental analysis.

If there is a consensus to continue the project, systems analysts then should perform the following activities to develop a functional specification for the proposed system:

1. Draft objectives of the proposed system.
2. Review the draft of objectives with all users.
3. Revise preliminary objectives accordingly.
4. Develop a logical system's design independent of physical devices, machinery, locations, and departments.

5. Identify what the users can and cannot expect from the proposed solutions.

6. Configure alternative physical solutions.

7. Identify and estimate the costs, benefits, limitations, exposure, and time frames associated with each physical solution.

8. Recommend which alternative physical solutions seems best for the particular environment.

9. Get a decision from the users on which physical solution to implement.

10. Revise the logical design of the proposed system in light of the physical alternative chosen.

11. Develop measurable statements of objectives in light of the physical alternative chosen.

12. Describe the proposed system, refining capabilities, responsibilities, impacts, and estimates of costs, benefits, and time frames.

13. Document all the logical terms and processes of the proposed system.

The preceding lists of activities and responsibilities take the systems development process to the detailed physical systems design phase. If we use only the narrow definition of systems analysis, both the analysis phase and the systems analyst's role have been completed. But, important questions remain to be answered:

- Where does that leave the users?

- Have their objectives been met?

- Who will evaluate the physical system's design to assure that it has not drifted from its originally defined purposes?

- Who will evaluate the implementation to assure that it is organized properly and that the system will be maintainable?

- Who will assure the users that the system has been tested properly?

- Who will see to it that people have been trained and are ready to use the system?

- Who will represent the users' interests until they are totally satisfied?

These tasks usually are the responsibility of the systems analyst.

If we use the broad definition of systems analysis, an important question still remains unanswered: Can anyone be knowledgeable enough to be a systems analyst and to do systems analysis? Within the field of medicine, there are internists, psychiatrists, heart specialists, and obstetricians, among others. Within the field of law, there are corporate lawyers, tax lawyers, criminal lawyers, estate lawyers, and so on. Is the data processing field now so large and so technically complicated that no one person can do the job of systems analysis?

Before answering that last question, let's look further at the increasingly complex problems the systems analyst is asked to solve.[8] A number of organizations now are undertaking projects so large and so intricate that one could say, in truth, that three or four systems are being developed concurrently. While an application system is being developed, for example, a data base system also may be developed to help support it. The application system may interface with a telecommunications network system. And the entire systems development effort may be so complex that a personnel system also may be needed to assure that all man/machine interactions are coordinated properly. Figure 1.5 illustrates role interactions.

It is fairly easy to see that the massive amount of technical knowledge required to develop complex systems goes beyond the immediate grasp of any one person. For this reason, we cannot expect the systems analyst to be competent in all of the technical areas required. The systems analyst, however, can solve these problems and evaluate solutions by systematically breaking down complex problems into more manageable subsets of problems.[9]

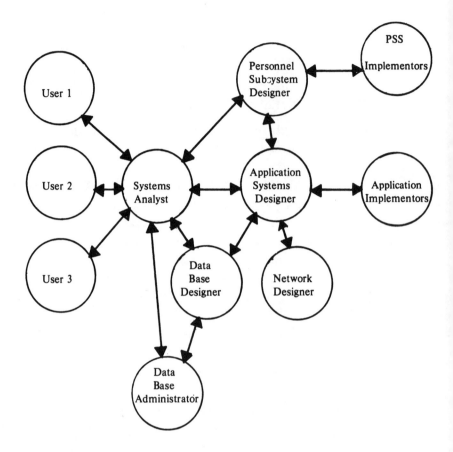

Figure 1.5. Growing complexity in systems development.

The systems analyst may not have technical expertise within a specific area, but he can draw upon the various specialists who do have that information. Of course, there is a danger in randomly selecting only part of the relevant information required for a solution, but the systems analyst can offset this by being aware of the problems and by compensating for them by drawing from available resources.

By better understanding the technical complexity of systems, we can better understand the role of the systems analyst. We can see that when a systems analyst comes upon a problem, he has to be able to analyze it in the context of the specific environment. He has to be able to determine a line of attack — whom to see, what information to gather, how to get that information. He has to be able to analyze that information to formulate and evaludte potential solutions. He has to be able to communicate these alternative solutions and follow through to assure that the objectives have been met in terms of schedule, budget, and systems requirements. The systems analyst does not necessarily have to know IMS, CICS, or VS. He doesn't have to be able to program in BAL, COBOL, or FORTRAN. He doesn't need to know everything there is to know about minicomputers or mass storage devices. *He does need to know whether this knowledge exists, where it exists, and how to get it, if needed.* He must be able to identify and communicate the relevant information. He must be a generalist, capable of distinguishing the forest from the trees. He has to have the skill to deal with the various levels and personalities of people found in a systems development environment. He has to have imagination in order to anticipate the problems that are likely to arise in a field where things are changing faster than ever before. He is a kind of explorer.

Keep in mind that the job *has* to be done. As problems and projects increase in scope and complexity, many more systems analysts may be required. They must possess analytical skills and a broad perspective to sort out conflicting demands in dynamic environments. Fragmenting the analyst's responsibility into specialized areas requiring specific expertise only will result in too many cooks in the kitchen, with none having the authority nor the understanding of the broad picture to be truly effective. Business analysts from user areas may bring specific knowledge of their special business requirements to a problem. Hardware specialists, data base specialists, and financial analysts may contribute their specific areas of expertise. But someone must take responsibility to assure that the objectives and priorities of all the users have been met. This burden is the responsibility of the systems analyst.

1.6 Organizational ramifications

Organizations have recognized that a broad scope of expertise is needed to provide continuity and to coordinate large systems development efforts. Regardless of how responsibilities for tasks are defined, the following are key activities in the systems development life cycle:

- identification of the problem
- definition of the current environment and its limitations
- determination of feasibility of alternatives
- definition of objectives and priorities
- preliminary design and comparative evaluation of alternative physical solutions
- definition of functional processing requirements
- design of physical system
- development of specifications for programs, modules, and procedures
- implementation of system
- optimization of system
- acceptance of system
- installation of system
- post-evaluation of system

The overall systems development effort could be a failure if any one of these key activities were skipped.

Organizations have expended much time and energy in determining the best organizational structure to deal with these key activities, but they frequently have been stymied. Training people in a number of areas, for example, is good, in theory; but, in practice, the effort is costly. It just is not realistic to think that a business analyst can learn completely about data processing or that a systems analyst can learn enough about a particular business in only one or two months' time.

Many organizations have found that the best way to deal with systems development problems is to set up flexible organizational structures that change with the specifics of the problem and the expertise of the personnel available; other organizations have established roles with fixed analytic responsibilities. Within some organizations, analysts report to management in the user areas. The essential role of these "business analysts" is to understand their user's business, so that they can define systems requirements to a systems development department. These business analysts stay with the project through its completion, and are, therefore, under considerable pressure to do a superior job (knowing that the implementation of their analysis could come back to haunt them). They follow through on their analysis by developing objective acceptance tests and by reviewing and verifying test results.

While business analysts do bring valuable business awareness and objectivity to the effort, they are somewhat remote from the actual systems development area. Frequently, they are unaware of the technological possibilities, do not understand the precision required for specifications, and are unable to evaluate physical designs. And, in cases in which the system serves multiple users (each with his own business analyst), the business analyst represents only a narrow, fragmented view of the overall system's objectives and requirements.

When analysts are in short supply and must be shared among a number of projects, organizations sometimes assign them to a project on a temporary basis to get the job started. The analyst's organizational responsibility, in this case, is much like that of a salesman for an engineering company: The "promoting" analyst asks the "customers" (the users) what they want, and specifies these requirements to the "manufacturing" department (in our situation, the designers and programmers). Typically, after the bulk of the analysis is completed, the promoting analyst transfers his main focus to another job, maintaining only a loose relationship with the first project. The main problems in this relationship are that the promoting analyst may not know much about the particular business application; his career path lies neither with systems development nor in the users' areas; and he may not be able to solve subsequent problems because of assignment to another project.

Most organizations have defined an organizational structure in which analysts report to the same functional area as do designers and programmers. These systems analysts may have responsibilities in project management, in logical design architecture, and even in physical design and programming. They usually bring to the job a broad perspective, analytic expertise, and a sound technical background. Drawbacks to this reporting relationship may come from the systems analysts' lack of an in-depth understanding of the business problems involved. In addition, their rapport with the systems development team may cloud their objectivity in conducting acceptance tests on behalf of the users and in criticizing the work of colleagues.

An important point about these individual reporting relationships is that they all have some weaknesses. As the problem, politics, and personnel change, so should the structure of the solution, but organizations usually cannot change their fundamental structure from project to project. As a result, existing roles and expertise sometimes do not match project requirements and responsibilities. For example, the systems analyst may not understand the objectives and requirements of the users; or the business analyst may be unable to evaluate the physical design of the system or to verify that the system had been properly tested; or the programmer may be responsible for developing the test data and accepting the test results.

Business analysts and systems analysts, however, each are reasonably effective in attacking parts of a systems development problem. In areas where a business analyst is weak, he can be helped by a systems analyst, and vice versa. (This collaboration is illustrated in Fig. 1.6.) In fact, it seems that the best analytic jobs are a result of a collaboration between business analysts and systems analysts. This approach involves the least training and provides the broadest scope of knowledge about the problem and its potential solutions. It assures that users and developers will communicate with each other, and that they will have the necessary broad perspective and objectivity to evaluate the solution effectively. It also leads to a minimum of surprises. When business and systems analysts pool resources and experience, personnel capabilities are properly matched to project responsibilities, and project coordination and continuity are enhanced. These should be the objectives of any organization involved in systems development.

	ability to understand the business	broad perspective	responsibility through completion	technical competence	ability to create functional specs	objectivity in acceptance	ability to communicate precisely	total
business analyst	3	1	4	2	2	3	2	17
systems analyst	2	3	2	3	3	1	4	18
promoting analyst	1	2	1	1	1	2	1	9
business and systems analyst	4	4	3	4	4	4	3	26

(scale = 1 to 4, worst to best)

Figure 1.6. Analysts' effectiveness comparison.

Review Exercise

1. Why have most systems analysts had very little, if any, training in how to do their jobs?

2. Why have most educational institutions excluded systems analysis from their EDP curriculum?

3. What is the distinction between the analysis phase and the analysis process?

4. What do we mean when we say that design involves the analysis process?

5. Why might a systems development effort involve many analysts from different areas?

6. What are the most important qualities or characteristics of a successful systems analyst?

7. What difficulties might be expected in a systems development effort if a systems analyst's responsibilities were to be completed at the end of the analysis phase?

8. Break down the systems development effort into at least five functional phases. Within each phase, identify two or more potential responsibilities of the systems analyst.

9. How can a systems analyst effectively handle complex systems development efforts requiring additional technical expertise?

10. What do we mean when we say that users are interested in the whole system, not in just a part of the process?

11. What are the strengths and weaknesses associated with business analysts, as compared to "promoting" analysts and systems analysts?

12. What are your ideas on the best reporting relationship between users, analysts, designers, and programmers?

References

1. D.L. Dance, "Is Anyone Here a Systems Analyst?" *Proceedings of the Fourteenth Annual Computer Personnel Research Conference,* ed. T. Willoughby (New York: Association for Computing Machinery, 1976), pp. 29-35.

2. E. Addleman, "There's No Such Thing as a Systems Analyst," Ibid., pp. 36-43.

3. Ibid.

4. P.W. Metzger, *Managing a Programming Project* (Englewood Cliffs, N.J.: Prentice-Hall, 1973).

5. W. Hartman, H. Matthes, and A. Proeme, *Management Information Systems Handbook* (New York: McGraw-Hill, 1968).

6. R. Ackoff, "Towards a System of System Concepts," *Management Science,* Vol. 17, No. 11 (July 1971), pp. 661-71.

7. W.S. Donelson, "Project Planning and Control," *Datamation,* Vol. 22, No. 6 (June 1976), pp. 73-80.

8. C. Alexander, *Notes on the Synthesis of Form* (Cambridge, Mass.: Harvard University Press, 1964).

9. E. Yourdon and L.L. Constantine, *Structured Design: Fundamentals of a Discipline of Computer Program and Systems Design,* 2nd ed. (New York: YOURDON Press, 1978).

2 The Evolution of Structured Analysis

The systems analyst travels light, usually with no more than a pencil, a template, and his intuition.

2.1 Perspective

Historically, EDP project development has resulted in over-budget, inflexible systems, which have not necessarily met the objectives and requirements of the users. As the complexity and variety of problems facing the systems analyst increase, it becomes painfully clear (at least to management) that traditional intuitive and trial-by-error approaches to problem-solving are inadequate. More than ever, the systems analyst needs a disciplined approach to solve problems in a cost-effective way.

Structured analysis is just such an approach. It is applicable throughout the entire systems development life cycle, but is especially effective in the analysis phase, when the definition of objectives, constraints, and requirements forms the system's foundation. It is structured in the sense that it utilizes a set of tools and guidelines that have proven successful in other fields, most notably structured programming and structured design. Let us look at how these disciplines have contributed to the concepts of structured analysis.

2.2 The success of structured programming

Structured first was used regularly in the data processing field to describe specific guidelines to be followed when writing a program — thus, structured programming.[1] Without specific

26

guidelines and techniques, each programmer developed his own haphazard strategy to program systems. There was no standardization from project to project or even from program to program. Some programmers coded and tested their programs twenty-five times faster than others. Some programmers produced code ten times faster in ten times less memory than others.[2] In general, though, programmer productivity figures were appalling. Programmers spent most of their time doing tasks other than programming,[3] and the programs they produced frequently were difficult to read and maintain.

In 1966, Böhm and Jacopini formally proved the basic theory of structured programming, that any program can be written using only three logical constructs,[4] described below and illustrated in Fig. 2.1.

- the process construct: a generalized, simple imperative statement mechanism (READ, ADD, MOVE)

- the decision construct: a generalized binary decision-making mechanism (IF-THEN-ELSE)

- the loop construct: a generalized looping mechanism (DO-WHILE, PERFORM-UNTIL)

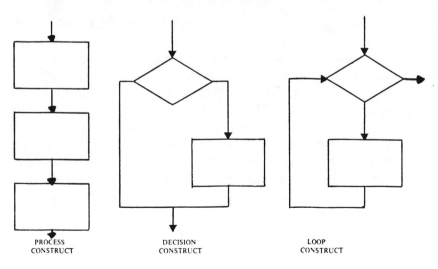

PROCESS CONSTRUCT   DECISION CONSTRUCT   LOOP CONSTRUCT

Figure 2.1. Constructs of structured programming.

In general, the techniques of structured programming have been dramatically successful in improving both the quality of code and programmer productivity. Typically, instead of the traditional average of ten to fifteen debugged lines of application code produced per person per day,[5] structured programming increases productivity to an average of thirty to forty lines of debugged code per day.[6] Moreover, the code usually is cleaner and more maintainable. The reason for this increased reliability and productivity is that structured code is more readable. Programmers can find and correct their bugs more easily, and there is a better chance that other parts of the system will not be affected when the code is changed.

2.3 The need for structured design

Unquestionably, structured programming has been successful, but it has not solved all of our problems. For example, when the code produced from structured programming is analyzed, flaws in systems design sometimes become quite obvious. Duplicate decision-making, excessive creation and passing of switches, and unmanageable modular structures are all symptoms of poorly structured code, which may be the inevitable result of poor systems design.

One could say that programmers should know better than to make the same logical decisions repetitively or to create and pass switches up and down the modular hierarchy. But in many organizations, programmers no longer see the entire picture of a system. They do their jobs knowing only the specifications of a particular module. When a specification says, Make this test and set on that switch when a particular condition is met, programmers do just that. How are they to know that the same test is being made in four other modules of the system? How are they to know that the switch is neither necessary nor desirable? The fact is that not only don't they know about such things, but they should not even care because these are considerations of systems design, not of programming.

Not only does critical analysis of code point out weaknesses in design, but also analysis of overall systems lifetime costs indicates that systems should be designed much more for maintainability than previously was done. In the 1950's and 1960's, machine costs associated with systems development exceeded people costs. Designers and programmers had to be preoccupied with program and machine efficiency. They were concerned, for example, with fitting a program in 4K of memory or executing a routine in 300 microseconds. Unfortunately, this preoccupation with space and time has carried over into today's environment in which people costs far exceed machine costs, [7] as shown in Fig. 2.2. The result is that too often designers and programmers alike overcomplicate the problem by developing sophisticated but unmaintainable solutions.

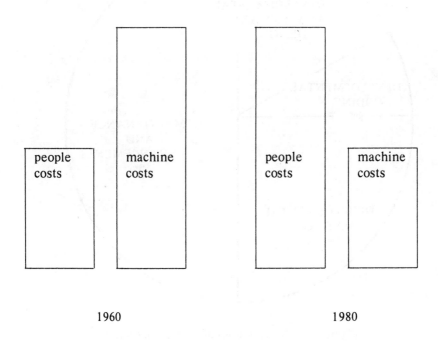

Figure 2.2. People and machine costs in perspective.

Another aspect of overall systems lifetime costs is that maintenance and enhancements consume typically at least 50 percent of the entire EDP budget in most organizations. As shown in Fig. 2.3, testing typically consumes about 50 percent of systems development costs (or about 25 percent of the total),

and we can see that developmental testing and systems maintenance usually consume at least 75 percent of overall systems
costs. This analysis points out that systems should be designed
so that they can be debugged and maintained easily. As a result,
there has been a revival of the ideas of Constantine et al. on
designing systems for lowest lifetime costs [8] and a recognition of
the value of structured design. [9,10]

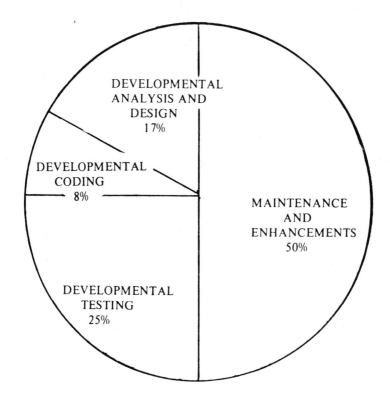

Figure 2.3. Systems lifetime cost breakdown.

Structured design prescribes specific guidelines to overcome
three basic problems of systems design — creation, evaluation,
and communication of the design document. Design creation is
based on analyzing the flow of data through a system. A logical
modeling graphic, the data flow diagram, is used to identify the
logical functions required to transform input data to output data.

These logical functions then are organized into a structure chart, a physical hierarchy of modules. Each module has one entry, one exit, and is designed to be as independent from other modules as is realistically possible (so that all of its elements or instructions will contribute to a single function). The result is a system composed of functional, independent modules. Such a system is not susceptible to the "ripple effect" in which a change to module A somehow causes a bug in module B, and so on. Obviously, this effect can, and too often does, become costly in consuming both human and machine resources.

Historically, systems analysts and designers have had difficulty in building a tangible model of a system that can be communicated to users. Typically, physical design solutions have been expressed in volumes of English narrative, sprinkled with technical jargon and supported by program-level flowcharts. The result has been a document that has overwhelmed users and obscured the logical requirements of systems. By contrast, an important aspect of structured design includes designing on a *logical* level before committing to the specifics of a particular *physical* solution. A logical design is easier to communicate because it is more concise, less technical, and uses graphics. Once the users have agreed to the logical requirements of the system, the logical model is transformed into a structure chart, the physical model of the modules in the system. By illustrating how the system's modular functions fit together, the structure chart serves much as does an architect's model of a proposed house or an engineer's scale model of a machine. The models of the system (the data flow diagram and the structure chart) are particularly powerful because they can be changed much more easily and cheaply than the resulting detailed code of the system.

2.4 The evolution of structured analysis

Structured design requires well-defined statements of inputs to the system, outputs from the system, data structures within the system, and processing logic that the system has to implement. These statements, unfortunately, must be derived from outdated documentation, and imprecise definitions of the problem and of the system's objectives. To improve our analytic methods, especially in the determination and communication of the needs, wants, objectives, and priorities of the users, we have looked for historical data in any field that could contribute. Studies of organizational structure, of effective management techniques, and of human behavior all have provided insights to many of the systems analysts' problems. We also have borrowed some of the techniques and approaches that have proved so successful in the areas of structured programming and structured design. The following important guidelines for analysis have been derived from these and other sources:

1. Get the users to participate in the development and evaluation of systems.

2. Consider the technical level of expertise and the bottom-line objectives of the users when producing documents for user review.

3. Use graphic tools to minimize potential communication problems.

4. Build a logical systems model before concentrating on detailed physical requirements.

5. Take a disciplined top-down approach to the analysis and design of systems. Break down major functions into manageably small, component functions.

6. Take a disciplined top-down approach to the implementation of systems. Address the implementation of major functions and the resolution of major potential problems before considering the more predictable detailed systems' functions.

7. Show the users output that they can understand before final systems acceptance. Break large, complex systems into smaller, more manageable projects so that users can see results early and can better evaluate the progress and value of systems.

8. Evaluate systems not only in terms of costs of development and operation, but also in terms of overall lifetime costs and benefits.

In a sense, structured analysis is *a disciplined approach to structuring the systems analyst's job.* Because of the variety of functions and of systems development phases in which the systems analyst is involved, structured analysis is defined as a *philosophical, top-down approach to all phases of the systems life cycle.* Where the problems are is where the systems analyst can be found. In systems development, the whole is not equal to the sum of its parts, but rather to the sum of its parts fitting together. As a result, structured analysis addresses not only the different phases of systems development in the form of a structured methodology, but also the communication and coordination between phases required to make that development a success.

In addition, structured analysis recognizes the need to look more closely at other problem-solving situations to identify tools and strategies that have proved successful in other fields. For example, we need to address the following questions: Why has our presentation format been reduced to distributing unreadable documents? Why don't we clearly define our terms as we use them, rather than document our systems after the fact? Why can't we trace data in a system like organizations trace paper flows among departments? Why can't we produce a blueprint of a system like an architect produces a scale model of a house? Why can't we review the completeness and structure of our logic before it has to be recompiled and retested? Why can't we produce flexible, maintainable systems? We can and we will — by adapting the strategies and tools of structured analysis.

The plumber, the carpenter, and the electrician come to a job with tools; but the systems analyst travels light, usually with no more than a pencil, a template, and his intuition. Structured analysis identifies a variety of graphic communication tools that depict situations better than words can express. Data flow diagrams and data structure diagrams, for example, provide logical models of application functions and of data base requirements that can be used as blueprints before a system is physically implemented. Organizational charts and structure charts illustrate the physical relationships of departments and of modules, respectively. Decision tables, decision tree structures, and structured English provide clear statements of policy and of program logic. A variety of graphics also can help in monitoring project progress and manpower loading. No matter what the situation, bar charts, line charts, circle charts, matrices, and tables can be used to emphasize and bring points into focus. Like the plumber, carpenter, and electrician, the systems analyst now has a bag of tools to help do the job.

Review Exercise

1. Why does structured analysis focus upon the analysis phase of the systems development life cycle?

2. What are some of the characteristics and benefits of structured programming?

3. What are the positive and negative aspects of showing a programmer only a small portion of the overall system's picture?

4. Why should maintenance be a strong consideration in the systems development perspective?

5. What is the value in building a logical model and a physical model of a system?

6. How does a logical systems model help in the systems development process?

7. What is meant by "the whole is not equal to the sum of its parts," and what is its relevance to a systems development environment?

8. What is structured analysis, and what purpose does it have?

9. What is structured about structured analysis?

10. Why might some people think of structured analysis as the science that never was nor ever can be?

References

1. O. Dahl, E. Dijkstra, and C. Hoare, *Structured Programming* (New York: Academic Press, 1972).

2. H. Sackman, W.J. Erikson, and E.E. Grant, "Exploratory Experimental Studies Comparing Online and Offline Programming Performance," *Communications of the ACM,* Vol. 11, No. 1 (1968), pp. 3-11.

3. G. Weinwurm, *On the Management of Computer Programming* (New York: Auerbach Publishers, 1970).

4. C. Böhm and G. Jacopini, "Flow Diagrams, Turing Machines and Languages with Only Two Formation Rules," *Communications of the ACM,* Vol. 9, No. 5 (1966), pp. 361-70.

5. F.P. Brooks, Jr., *The Mythical Man-Month* (New York: John Wiley & Sons, 1975).

6. E. Yourdon, *Techniques of Program Structure and Design* (Englewood Cliffs, N.J.: Prentice-Hall, 1975).

7. G. Amhdahl, "Large Commercial Systems, Their Architectural and Technological Evolution," presented at the Sixth Annual N.Y. State Government Data Processing Conference, June 1976.

8. L. Constantine, "Modular Programming," *Proceedings of a National Symposium,* ed. T.O. Barnett (Cambridge, Mass.: Information & Systems Press, 1968).

9. W. Stevens, G.J. Myers, and L.L. Constantine, "Structured Design," *IBM Systems Journal,* Vol. 13, No. 2 (1974), pp. 115-39.

10. E. Yourdon and L.L. Constantine, *Structured Design: Fundamentals of a Discipline of Computer Program and Systems Design,* 2nd ed. (New York: YOURDON Press, 1978).

PART 2

Tools of Structured Analysis

3 Introduction to the Tools of Structured Analysis

I could not solve the problem merely by thinking about it and by applying a haphazard, intuitive approach. I did not recognize the need for resources beyond those I had been using. I needed an analytic tool to help me structure and picture the problem.

3.1 The hat problem

In an introduction to the tools of structured analysis, the tools could be identified and their value emphasized. But would that sell you on their merit? If, instead, the power of the tools could be made apparent, would you not be more convinced of their value? Allow me, then, to tell you about an incident that occurred during my university days, when I was a student in a logic course. One day, the class was given an exercise. The problem did not seem too difficult; indeed, it seemed so simple that I kept thinking that I was only about one minute away from getting the answer whenever the problem came to my mind. I was of course wrong, because the problem came to my mind many times; and after spending hours thinking about it, I was no closer to the answer than when I had started. In fact, I came to the conclusion that if I could not solve this seemingly trivial problem, there had to be something wrong with the problem. (It's not my program; it must be the machine.) So a few days later when the class met again, I announced there was insufficient data to solve the problem. Sure enough, within five minutes I was proved wrong again, because an irrefutable solution was presented. The problem is as follows:

There were three prisoners (A, B, and C) in jail. One day, the jail-keeper decided to have some fun with the prisoners. He said, "I'm going to put blindfolds on each of you. Then, I'm going to put a hat on each of your heads. I've got three red hats and two white hats. One by one, I'm going to take the blindfolds off. You will be able to see the hats worn by your jail-mates, but you will not be able to see the color of your own hat. If you can tell me what color hat you are wearing by looking at the hats worn by the other two, I'll let you out of jail. But if any of you guesses wrong, I'll put all three of you in solitary confinement for the next two months."

Having explained the rules, the jail-keeper put the blindfolds and hats in place. The blindfold was removed from prisoner A, who was asked what color hat he was wearing. By looking at the color of the hats worn by B and C, A could not make a determination and said, "I don't know." The blindfold was removed from prisoner B, and B was asked what color hat he was wearing. By looking at the color of the hats worn by A and C and with the information he had gained from A's answer, B could not make a determination and said, "I don't know." As the blindfold was removed from C, the jail-keeper laughed perversely — for C was a blind man. But C said, "Don't laugh at me. Although I do not have sight, I clearly know from what my friends with eyes have said that my hat is _____." C was correct, let out of jail, and lived happily ever after.

The questions are, What color hat was C wearing and how did he know its color? Your only clue is that A, B, and C all were intelligent and responded in turn with the best possible answers available to them.

Before reading further, you are invited to try to solve the hat problem. Looking back at what happened to me in that logic class, I recall my inability to focus on the combinations of conditions in the problem. It seemed so simple that I felt sure I could solve it merely by trying. But, as I found out, the complexity of the problem extended just beyond the limits of my intuition. As I mentally moved from one set of conditions to another, the problem wasn't so much that I kept forgetting the sets of conditions that I had eliminated; instead, I could not picture the sets of conditions that I had not eliminated, and their relationship to the question at hand. I could not solve the problem merely by thinking about it and by applying a haphazard, intuitive approach. I did not recognize the need for resources beyond those I had been using. I needed an analytic tool to help me structure and picture the problem. Such a tool is shown in Table 3.1, below.

Table 3.1
Decision Table for Hat Problem

CASE	PERSON			ACTION
	A	B	C	
1	W	W	W	
2	W	W	R	
3	W	R	W	
4	W	R	R	
5	R	W	W	
6	R	W	R	
7	R	R	W	
8	R	R	R	

A decision table, such as the one shown in Table 3.1, would have been very helpful in picturing the various combinations of conditions in the problem. From it, we see that Case 1 can be eliminated because there are not three white hats in the problem. Cases 3 and 5 can be eliminated because both A and B would have been smart enough to know that they each were wearing a red hat if either of them had seen two white hats. I suppose that anybody who has ever attempted to solve this problem has gotten at least to this point. However, here is where most of us go astray, because we do not comprehend what has yet to be analyzed and which are the critical cases.

Five combinations remain to be analyzed. In fact, however, a close look at the decision table will indicate that only one more combination has to be eliminated in order to solve the problem. Cases 2, 4, 6, and 8 all result in C having a red hat; only Case 7 results in C having a white hat. If we somehow can eliminate Case 7, we will know that C was wearing a red hat and understand how C figured it out. The decision table has provided us with a picture so that we can see where we should be spending our time — not on Cases 2, 4, 6, and 8 (which, by the way, cannot be eliminated), but rather only on Case 7.

How to eliminate Case 7 certainly is not as trivial or obvious as the elimination of Cases 1, 3, and 5. Case 7, however, can be eliminated by applying the following reasoning: If A had seen a red hat on B and a white hat on C, A would have been unable to identify the color of his own hat. A's inability to answer the question definitively, however, would have contributed the information B would need to answer the question. If B had seen a white hat on C, B would have known that he had to be wearing a red hat. (If C had a white hat, and if A could not tell his hat was red, A would have seen a red hat on B.) In that B did not know his hat was red, C could not be wearing a white hat, and Case 7 is eliminated. By going back to the decision table and adding remarks in the Action column, we can picture the solution more clearly. (See the remarks in the Action column in Table 3.2, shown on the following page.)

It is unlikely that the specifics of the problem or its answer will help us in our everyday lives. But the problem helped to expose weaknesses in my problem-solving attitudes and techniques, and the importance of the lessons learned have impressed me to this day.

Table 3.2
Decision Table for Hat Solution

CASE	PERSON			ACTION
	A	B	C	
1	W	W	W	impossible — only 2 white hats in problem
2	W	W	R	
3	W	R	W	B would have known his hat is red
4	W	R	R	
5	R	W	W	A would have known his hat is red
6	R	W	R	
7	R	R	W	B would have known his hat is red
8	R	R	R	

Sometimes, you need a specific tool to do a particular job. In other situations, any one of many tools will do. The example of the prisoners with the red and white hats can be attacked using a decision table, a decision tree structure, or even systematic verbal documentation. They equally would provide the focus required to help solve the problem. Learning how to use these tools is not a difficult task. It is more difficult to recognize that a situation calls for analytic tools and, at that point, to determine which tools should be applied. Keep in mind, however, that each such tool is a means to an end, not the end itself. Using a tool helps in the organization of information, but does not actually solve the problem.

3.2 A hypothetical situation

Now, let us look briefly at a potential systems development situation and some of the tools that can aid the systems analyst in his job.

A mail order company advertises products in magazines. Most orders are initiated by magazine subscribers who fill in and send coupons to the mail order company. The company also takes orders over the phone, answers inquiries about products, and handles payments and cancellations of orders. Products that have been ordered are sent either directly to the customer or to regional offices of the company, which then handle the required distribution. The mail order company has three basic data files that retain customer mailing information, product inventory information, and billing information based upon invoice number. During the next few years, the company expects to become a multimillion-dollar operation. Recognizing the need to computerize much of the mail order business, the company has begun the process by calling in a systems analyst.

Let us assume that the systems analyst has identified the user community and has spoken to people in the various departments to determine the status and limitations of the current system. Let us also assume that the analyst has gathered all of the written documentation and periodic reports that describe the current system's objectives, operations, and performance. Having done all that, the systems analyst still may not have a clear picture of the current system's functions and its logical flow of data. What the analyst needs is a tool to help him build a logical model of the current system.

The data flow diagram (DFD), which is discussed in greater detail in Chapter 4, is just such a tool. It identifies the logical inputs and outputs of a system, their sources and destinations, and the logical processes required to transform the inputs to outputs. Figure 3.1 shows a logical data flow diagram, depicting the mail order company's current functional requirements. Were we to assume that the proposed mail order processing system would add no new functions, this logical DFD would picture both the current and the proposed systems' functional requirements, independent of physical implementation considerations.

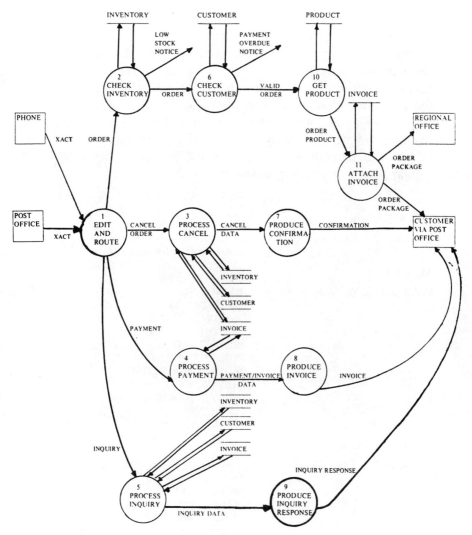

Figure 3.1. Logical DFD for mail order company.

A data flow diagram provides a *general* picture of the data transformation in a system. It does not describe the details of the data flows and processes that may have been identified by a systems analyst in meetings with users. Those details, even if they are incomplete and exist only on a logical level, should not be lost, and the analyst should assure that they are documented in a data dictionary. (See Chapter 8 for further information.)

To better understand this concept of documentation, let us return to our mail order example: Let us suppose that the analyst asked someone in the accounting department, "What information do you carry on the INVOICE file?" Whereas, it may be necessary ultimately to define this information at the bit level for the proposed system, in early analysis the physical characteristics of the data in the proposed system should not be a consideration. Instead, the systems analyst should define the logical characteristics of the current system's INVOICE file as a dictionary entry similar to the one shown below in Fig. 3.2:

FILE or DATA BASE NAME:	INVOICE
CONCISE DESCRIPTION:	Information by invoice number on invoices in the current system.
ALIASES:	NONE
COMPOSITION:	INVOICE-NO
	INVOICE-DATE
	CUSTOMER-NAME
	CUSTOMER-ADDR
	PRODUCT-NO
	PRODUCT-QUANTITY
	PRODUCT-UNIT-COST
	SHIPPING-COST
	DISCOUNT-RATE
	INVOICE-NET
	INVOICE-STATUS
ORGANIZATION:	Sequential by invoice number
ASSOCIATED PROCESSES:	from Overview DFD of Current System
	PROCESS-CANCEL(3)
	PROCESS-PAYMENT(4)
	PROCESS-INQUIRY(5)
	ATTACH-INVOICE(11)

Figure 3.2. Data dictionary entry for mail order company.

Our system may have to respond to all kinds of inquiries. The types and volume of inquiries and the required response time to inquiries may necessitate a proposal for a system with a complex data base. If so, the systems analyst should determine the logical data structure requirements before a physical data base is created. He must ask such questions as: To what kinds of inquiries must the system respond? How fast is the information needed? What special security or privacy considerations exist? From responses to these questions, the analyst should draw a diagram to picture the logical data relationships in the system, independent of how the information ultimately will be held or retained. This diagram should be sufficiently elementary to enable users to verify their data structure requirements, yet sufficiently detailed to provide the physical data base designer with a picture of the users' immediate-response requirements.

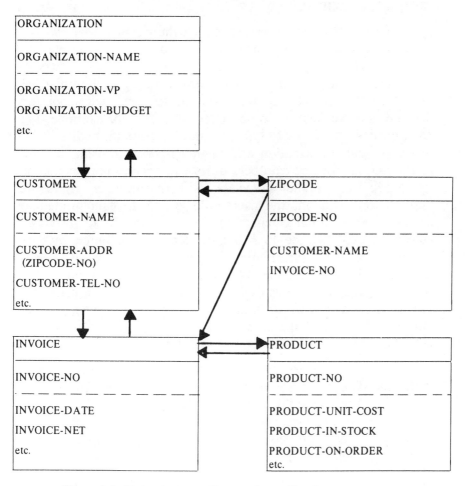

Figure 3.3. Data structure diagram for mail order company.

The data structure diagram (DSD) shown on the previous page in Fig. 3.3 indicates the logical data structure relationship requirements for the proposed mail order company system. We will deal with DSDs more extensively in Chapter 5, but we can learn several things from the data structure diagram depicted here. The diagram makes no definitive statement as to the ultimate physical structure of the data base. However, it does show that the users and systems analyst envision five data entities in the proposed system: ORGANIZATION, CUSTOMER, INVOICE, PRODUCT, and ZIPCODE.

Each entity is comprised of attributes that identify and describe sets of information within an entity. For example, the ORGANIZATION-BUDGET amount and the ORGANIZATION-VP name are properties of a particular organization identified by ORGANIZATION-NAME. Between the entities are logical pointers that indicate the relationship between entities. For example, Fig. 3.3 indicates that by identifying a product, users immediately can access product information and related invoice information. By identifying an invoice number, users can get invoice, customer, and product information associated with that invoice. By identifying an organization, the users will be able to gain immediate access to organization or customer information. By identifying a customer, users will be able to gain access to customer, invoice, and organizational information. Note that the logical pointer between the INVOICE and ZIPCODE entities is in only one direction. The diagram indicates that by identifying a ZIP code, the invoice information within that ZIP code can be retrieved quickly. However, by identifying an invoice, the user will not be able to access quickly the information in the ZIPCODE entity.

In virtually every data processing system, there are logical rules that determine how inputs are to be transformed. As the logical rules increase in complexity, English narrative descriptions become less acceptable as the specification tool. Three specification tools are suggested by structured analysis to support or replace potentially ambiguous and verbose English narratives: structured English, decision tables, and decision tree structures. To show their uses, let us assume that the analyst wanted to clarify the specific logic involved in confirming an invoice in the mail order company.

Figure 3.4, below, shows the logic's specification using structured English, and utilizes indention and parentheses to enhance readability and, ultimately, maintainability.

```
IF the invoice exceeds $500
|       IF the account has any invoice more than 60 days overdue
|       ¦ hold the confirmation pending resolution of the debt
|       ELSE (account is in good standing)
|           issue confirmation and invoice              .
ELSE (invoice $500 or less)
|       IF the account has any invoice more than 60 days overdue
|       ¦ issue confirmation, invoice, and write message on credit action report
|       ELSE (account is in good standing)
|           issue confirmation and invoice
END-IF
```

Figure 3.4. Structured English for mail order company invoice procedure.

Table 3.3
Decision Table for Mail Order Company Invoice Procedure

	1	2	3	4
INVOICE-AMOUNT	>$500.00	>$500.00	≤$500.00	≤$500.00
INVOICE-STATUS	OK	OVERDUE	OK	OVERDUE
HOLD CONFIRMATION		X		
ISSUE CONFIRMATION	X		X	X
ISSUE INVOICE	X		X	X
WRITE MESSAGE				X

Table 3.3 shows the same logic in the form of a decision table, providing a better picture with fewer words.

Figure 3.5 shows the same logic in the form of a decision tree structure, which has the advantages of being easy to create and to understand. All of these tools are described in greater detail in Chapter 6.

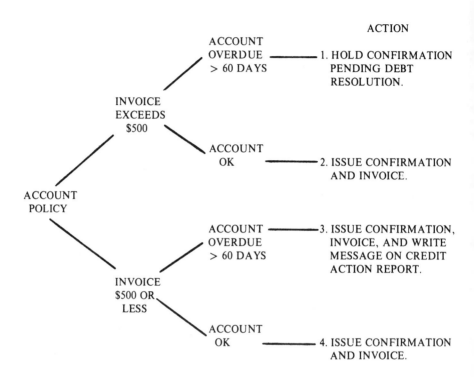

Figure 3.5. Decision tree structure for order company.

Logical designs in the form of data flow diagrams usually are converted into physical models of systems by two specific top-down design strategies: transform analysis, and/or transaction analysis (see Chapter 9). The resulting diagram is a structure chart showing the hierarchy of modules in a program or in a system, and the data communication required for the modules to do their functions. Figure 3.6, on the following page, illustrates a simple structure chart for a program that edits data. The main module gets sequenced input or an end-of-file indicator. It determines what kind of record has to be edited and passes control to a module to process a record type. The main module then receives an error flag indicating the results of the edit-processing, and either passes control to a module to write a valid record or returns control to get another sequenced record. In Chapter 7, we will expand this description of structure charts, but their usefulness can be seen from this early example.

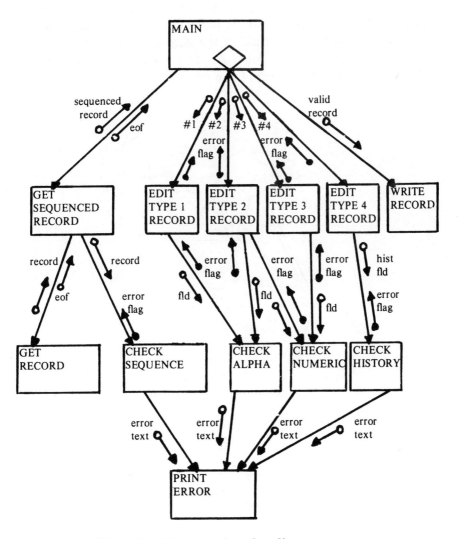

Figure 3.6. Structure chart for edit program.

3.3 The relationship among tools

The relationships among the tools of structured analysis are depicted in Fig. 3.7. The *data flow diagram* provides a logical model of the system, and depicts data flow and logical processes that are documented in a *data dictionary*. The data flow diagram may show that a logical process requires access to a data base. Data base requirements for immediate responses in the form of entity and attribute definitions are pictured in the *data structure diagram* and are documented in the *data dictionary*. Ultimately, a physical systems design must be developed from the data flow diagram, the logical model of the system.

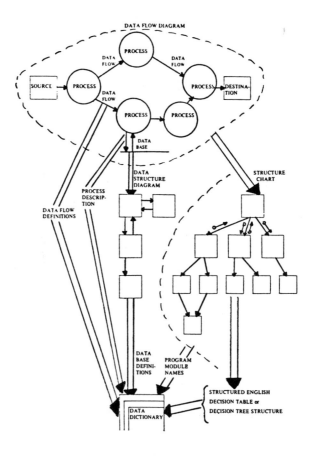

Figure 3.7. Relationship among tools of structured analysis.

Physical data definitions refine logical data flows from the data flow diagram and are documented in a data dictionary. Logical processes in the data flow diagram are transformed into a *structure chart,* which provides a physical hierarchy of modules to perform the logical functions identified. Program, module, and subroutine names are documented in the data dictionary, along with detailed module specifications in the form of structured English, decision tables, or decision tree structures. In this way, both logical and physical models of the system are developed, and the details of these graphics provide documentation *as the system is being developed.*

Review Exercise

1.   What lessons for systems analysis can be derived from the hat problem?

2.   Describe your own experiences in which you could and should have used tools to help solve a problem. What was the effect of not using tools in these situations? Why were these tools not used?

3.   If you have not used the tools of structured analysis in your analytic work, what tools *have* you used? How effective have they been? How do they compare to the tools presented in this chapter?

4.   Describe the problems you have had in developing or implementing systems or program specifications. Identify the graphics used in these specifications and any other graphics that may have been useful, but were missing from these specifications.

5.   Why has documentation frequently been done after the fact, and what are typical results of such a procedure?

4 Data Flow Diagrams

Actually, many systems are created without a clear understanding of the current system or of the logical requirements for the proposed system.

4.1 Perspective

As discussed in the previous chapter, the data flow diagram is a graphic tool used to depict the logical flow of data through a program or a system.[1] The DFD, also known as a *bubble chart* or *data flow graph,* is particularly useful in identifying the functions of a system and the resultant data transformations. The tool first was used before the computer era to show the paper flow in manual processes and the product development flow in manufacturing applications. Today, in a systems development environment, the data flow diagram can be used to emphasize the logical flow of data in a system, while de-emphasizing procedural aspects of the problem and physical solutions.[2]

4.2 Data flow diagram symbols

The basic symbols of the data flow diagram are called transforms; these are represented by circles, each identifying a function that transforms data. The circles are connected by labeled arrows, which represent the inputs to and the outputs from the transforms. Figure 4.1 shows these elements of a data flow diagram in abstract form. The diagram indicates a transform F that converts a stream of data elements known as X into a stream of data elements known as Y. Y subsequently is input into transform G, which produces data in the form of Z. Note that the diagram does *not* indicate whether one X produces one Y, or that one X produces many Ys, and so forth.

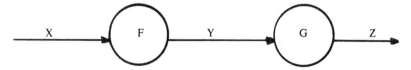

Figure 4.1. Basic symbols of a DFD.

Sometimes, transforms require more than one input and produce more than one output. When multiple data streams are either inputs to or outputs from a transform, the procedural association between the transform and its inputs and/or outputs may be unclear. For example, if three data streams are input to a transform, we may not know whether all three, two of three, or only one of three inputs is required for the transform to do its job. Systems analysts who use data flow diagrams have mixed opinions on whether the procedural association between a transform and its data streams should be shown on a data flow diagram. Some argue that additional notation to clarify the association of data streams only will clutter the diagram. They further point out that these are control details as opposed to data flow transformation details and, as a result, should not be part of a data flow diagram.

As we will see later in this chapter, data flow diagrams can be done in levels; the first data flow diagram created may be an overview of the system depicting general functions; subsequent levels may define the detailed functions required within a general function. The procedural association of data streams is a detailed control consideration, which normally will not add much meaningful information to an overview data flow diagram. Such information, however, may be useful to depict input-output control information within detailed functional processing. As a result, we generally should ignore data flow control considerations when developing overview data flow diagrams, but perhaps should include details of data flow control on lower-level, more detailed diagrams. Above all, we should try to consider who will receive these documents and what level of detail they seek. When the situation seems to call for detailed definition of the procedural association of data flows, the following notation is useful to define data stream relationships.

*   denotes a logical AND connection

⊕   denotes an exclusive OR connection

0   denotes an inclusive OR connection

Figure 4.2 shows both the logical AND and exclusive OR connections. Both on-line input and batch input are required for the EDIT INPUT transform. Neither input by itself is sufficient for the EDIT INPUT process. The figure also shows that as a result of the EDIT INPUT process, either a rejected transaction or a valid transaction (but not both) will be produced.

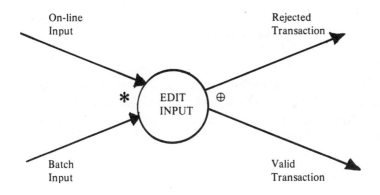

Figure 4.2. AND and exclusive OR connections.

The logical association of data streams is not always as simple as that shown in the previous figure. Various combinations of inputs or outputs may be required in a process. Figure 4.3 shows an UPDATE MASTER transform that requires master file input and some combination of transaction input. We can imagine the UPDATE MASTER process using the master file and transactions A, B, and C. We also can imagine the process using the master file and transactions A and B or transactions B and C. While the inclusive OR connector may not make the logical association of data streams absolutely clear, it does indicate a complex set of conditions that normally would be more clearly defined in other documentation, such as in a detailed write-up of the UPDATE MASTER transform.

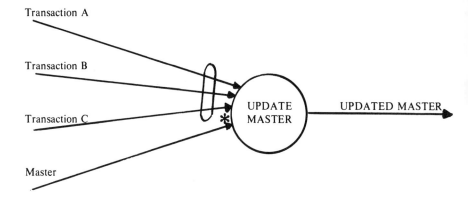

Figure 4.3. Inclusive OR connection.

Sometimes, a transform produces a set of data flows that can be represented only by use of mixed AND and OR connectors. For example, the output of an UPDATE MASTER transform may be both the updated master record and the updated history record (when the transaction is acceptable), or a transaction reject identification (when the transaction is unacceptable).

This situation can be pictured in a data flow diagram with the logical connectors just defined, but their use may leave reviewers with too much room for misinterpretation. (See Figs. 4.4a and 4.4b, on the facing page.) The ambiguity resulting from the use of mixed AND and OR connectors frequently can be overcome by splitting a data flow into multiple data flows, as shown at the bottom of the next page in Fig. 4.4c. Note that the three examples in the figure are alternative ways of showing the same structure. However, Fig. 4.4c is the only example to depict the processing requirements with no ambiguity.

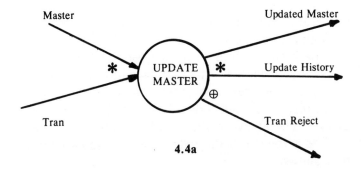

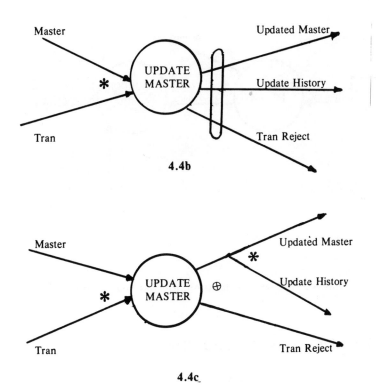

Figure 4.4. Mixed AND and OR connections.

Data streams come from a source and go to a destination. The sources and destinations of data streams are depicted graphically by rectangles, as shown in Fig. 4.5. The figure also shows access to a data base in order to do the required processing. For the purposes of a data flow diagram, a data base may be considered simply as data retained from transaction to transaction. Access to and response from a data base are depicted by a pair of labeled arrows: The arrow to the data base represents the search argument. The arrow from the data base represents retrieved information or status.

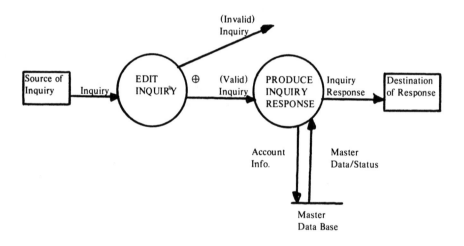

Figure 4.5. Source, destination, and data base representation.

4.3 Filtering versus transformation of data

Looking at Fig. 4.5, one might ask, What is the data transformation associated with the EDIT INQUIRY function? An inquiry is input to the function, and that same inquiry is output from the function. If we accept that the EDIT INQUIRY function does no modification of data such as padding, truncating, or blank-filling erroneous fields, then the input and outputs physically are the same. So, where is the transformation of data?

The fact is that there may not be any physical transformation of data. In the example noted above, the transformation of
data exists only in the sense that our *knowledge* of the data has
been transformed. We have made a logical refinement. We
know that the input is a raw inquiry. We know that the output
of the EDIT INQUIRY transform is either a valid inquiry or an invalid inquiry. Logically, there is quite a difference. In this case,
the data flow has been filtered as a result of the EDIT INQUIRY
transform. In another example, an unidentified transaction could
be input to an IDENTIFY TRANSACTION transform, which would
identify and distribute the transaction according to its transaction
type (see Fig. 4.6). Once again, the physical data are not
transformed; it is our *understanding* of the data that is enhanced
as a result of the process. Transforms, which represent functions, therefore can act as "filters" or "distributors," instead of
physically transforming the data.

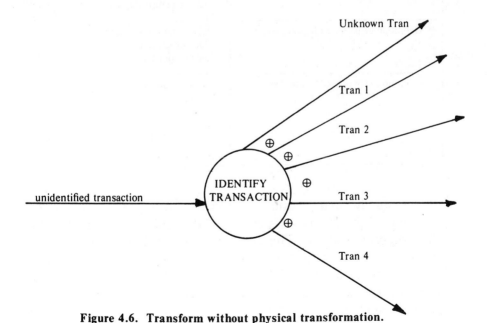

Figure 4.6. Transform without physical transformation.

4.4 Usage of data flow diagrams

In developing systems, we should identify *what* we currently are doing and *what* we would like the new system to do, before we determine *how* to solve the problem. Stated differently, the implementation of a system should proceed only after the steps presented in the table below have been documented:

Table 4.1
Analysis and Design Steps and Tools

STEP	TOOLS
Definition of current physical system	physical data flow diagram I/O flowchart
Definition of current logical system	logical data flow diagram data structure diagram
Definition of proposed logical system	logical data flow diagram data structure diagram data dictionary
Definition of proposed physical system	physical data flow diagram structure chart decision tables decision tree structures structured English data dictionary

Actually, many systems are created without a clear understanding of the current system or of the logical requirements for the proposed system. The current system's definition usually is inadequate — because current documentation often is out of date or incomplete, because the systems analyst may not be aware of tools with which to derive a current system's definition, and because systems developers are in a hurry to work on the proposed system. The logical requirements of the proposed system usually are not defined properly — because systems developers do not realize this critical need, and, again, because they are in a hurry and do not have the proper tools to do the job.

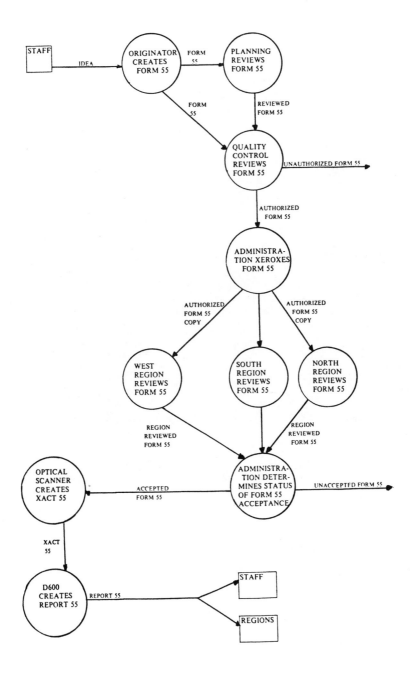

Figure 4.7. Physical DFD — geographic change processing.

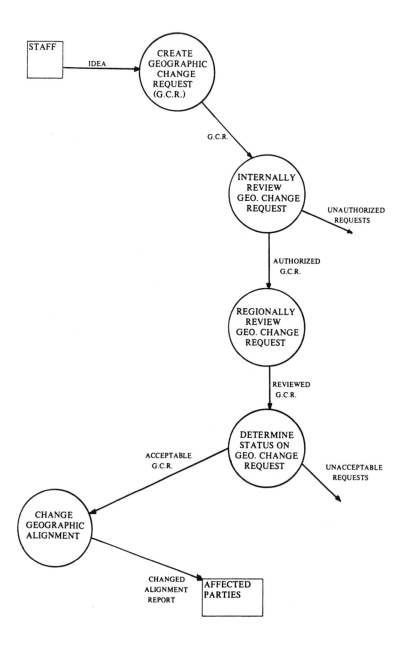

Figure 4.8. Logical DFD showing geographic change processing.

The data flow diagram is a tool that analysts can use to define both the current system and the proposed logical system. The current physical system can be depicted with a data flow diagram that identifies specific people, departments, physical locations, physical documents, and physical devices. For example, our mail order company that we discussed in Chapter 3 could have an existing set of procedures established to process a change in geographical alignment. These existing physical procedures and resultant data transformations could be pictured in a data flow diagram, as shown previously in Fig. 4.7.

A possible result of the study of the mail order company is that the company could be reorganized to eliminate regional offices. Similarly, within the central office, departments could be combined, discontinued, or in some way reorganized. The physical layout of forms and the physical devices in the system also could change. With these possibilities in mind, the systems analyst should view the current processing requirements as being independent of physical departments, regions, forms, and devices to identify the logical processes and data flows in the system. The analyst can do this by revising the physical data flow diagram of the current system so that it reflects no specific physical departments, regions, forms, or devices. The result, shown on the previous page in Fig. 4.8, is a logical data flow diagram of the current system. This logical diagram, with all physical terminology removed, provides users and nontechnical reviewers of the system with a document that they can understand; they do not have to know anything about Form 55, for example, the optical scanner, or the D600 to review the logical requirements of the current system.

Once a logical DFD of the current system has been created, the systems analyst should identify and communicate the logical changes in processing requirements called for by the proposed system. Frequently, the proposed system modifies or enhances current logical processing requirements. For example, for our mail order company, the systems analyst may determine that the proposed system will do everything functionally that the current system is doing, with the following additions and revisions:

- A record will be kept on all geographic change requests; the time, date, and source for every change request must be logged and filed.

- All geographic change requests will be reviewed and distributed only by the geographic coordinator in the central office.

- A historical record will be kept on all geographic change requests that are approved and that result in changes. This information will be appended to the record kept on geographic change requests.

Just as the physical data flow diagram of the current system could be used to help derive its logical data flow diagram, the logical DFD of the current system can be used to help derive the proposed logical system, as shown in Fig. 4.9 on the next page. Note that the logical requirements of the proposed system do not differ greatly from those of the current system. A data base has been added, a few functions have been added, and some functions have been eliminated.

4.5 Levels of data flow diagrams

Students who are developing data flow diagrams often ask, "How do we know when we are done? To what level of detail should we go in identifying functions?" Their questions are well-founded. One conceivably could develop a data flow diagram with only one transform, depicting a rather complex system (see Fig. 4.10). While such a diagram could be useful in identifying inputs, outputs, sources, and destinations, it does not identify any of the basic processes required to transform the inputs to outputs.

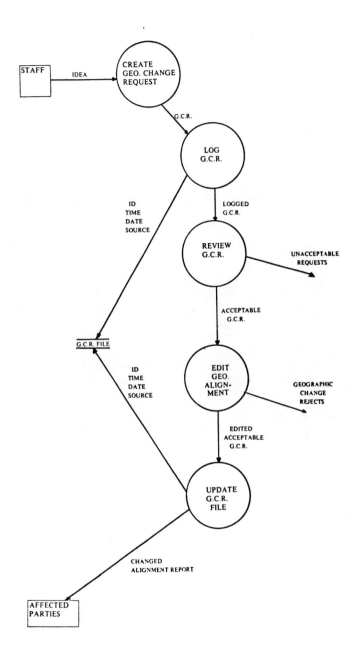

Figure 4.9. Proposed system logical DFD — geographic change processing.

Figure 4.10. A one-bubble data flow diagram.

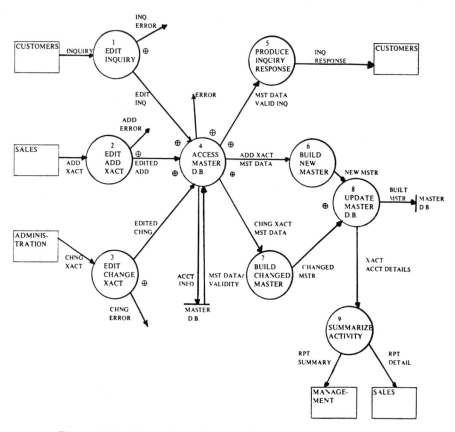

Figure 4.11. Generalized (overview) data flow diagram.

Figure 4.11 depicts the system shown in Fig. 4.10, but in greater detail. The diagram has considerable value in showing the basic functions and the data base required for the system. But do we know enough about the details of the problem from this diagram? By employing some of the strategies of structured design, can we envision each of the transforms as being functionally independent and, when implemented, manageably small? Figure 4.11 represents a good start at picturing the problem, but shows only a generalized overview of the data flow and functions of the system: Error paths are not resolved; reading and writing of records are not shown; the various types of editing requirements are not depicted. To show that information, we must expand each of the transforms into a more detailed DFD.

The overview DFD is used to show on one page the major processes required in a system. These processes normally must be broken down to show the detailed processing requirements within each major function, a task best accomplished by drawing levels of DFDs. To understand the detail of a given process, we should draw or refer to a DFD with the same numerical prefix as the process in question. Each process on the "parent" diagram is a consolidation of the network shown on the "child" diagram. Inputs and outputs are matched between parent and child, except for errors that may not be present on the parent diagram. (See Fig. 4.12 on the next page.)

Looking back at Fig. 4.11, we might envision that the EDIT CHANGE XACT transform is not as simple as it appears. There may be many different types of change transactions, each with specific editing requirements. To clarify the detailed functional requirements within the general function of change transaction editing, we would draw a more detailed, second-level data flow diagram, as shown in Fig. 4.13. Note that multilevel data flow diagrams facilitate top-down analysis, as major functions are decomposed into their subfunctions. In addition, the approach minimizes the need for off-page connectors and simplifies the process of updating or otherwise revising diagrams.

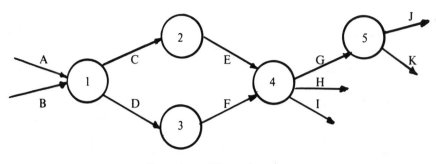

Overview - Figure 0

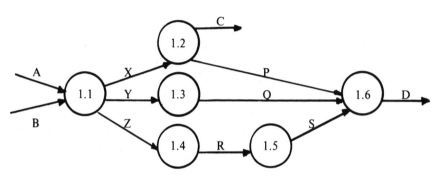

Second level - Figure 1
(Blowup of process 1)

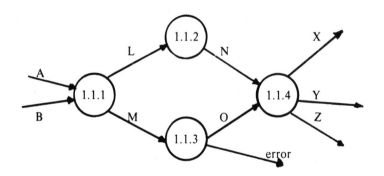

Third Level — Figure 1.1
(Blowup of process 1.1)

Figure 4.12. Levels of data flow diagrams.

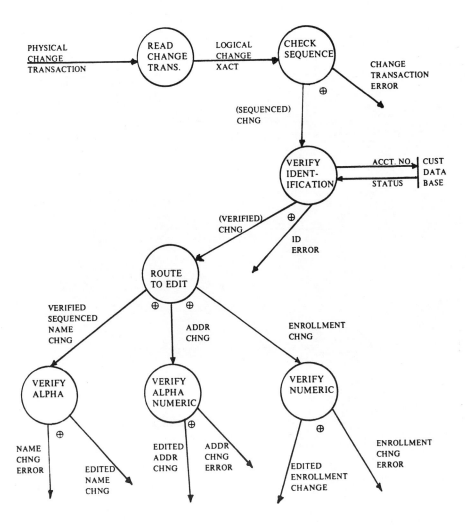

Figure 4.13. Second-level DFD with EDIT CHANGE XACT.

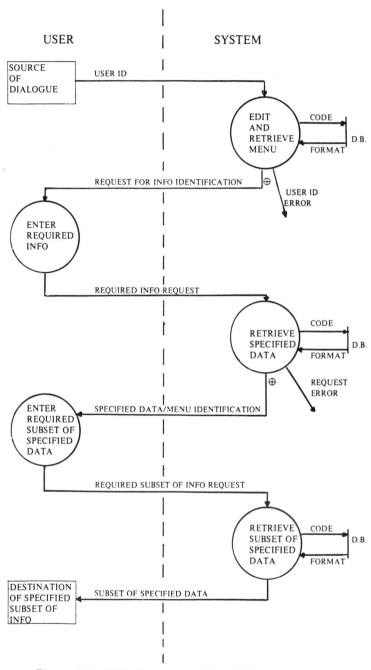

Figure 4.14. DFD for man/machine dialogue.

4.6 Data flow diagram variations

Data flow diagrams compress functional requirements and the resulting data transformations into a compact picture of the problem. In systems in which the user and the machine are in conversational mode, the data flow diagram can picture both activities. For example, Fig. 4.14, a nonstandard variation of the data flow diagram, shows all user functions on one side of the page, and all machine functions on the other. The responsibility for the function clearly is indicated by the dotted line in the center of the diagram.

A similar technique can be used to show departmental responsibilities in a physical data flow diagram. Referring to Fig. 4.7, note that each transform has a subject, verb, and object. The subject is the department, region, or device that is responsible for the function identified by the verb and object in each transform. Alternatively, we could have pictured the situation by identifying the subjects at the top of the diagram, as in Fig. 4.15. Functions associated with each subject are identified by their vertical positioning under each subject. This type of diagram is particularly useful when an analyst wants to verify a specific department's interaction with the system. Although that department may neither know nor care about the interaction of other departments in the system, the diagram does provide that information. In addition, the DFD provides a clear picture, by department, and a focus for departmental review.

Recognizing the basic need for a tool such as the data flow diagram, some organizations have developed or adapted similar diagraming techniques to show the logical inputs, outputs, and processes in a system.[3] While most differences between data flow diagrams and other diagraming techniques relate to style rather than substance (rectangles instead of circles, for example), Sof-Tech has developed the Activity Diagram, which depicts control requirements as well as inputs, outputs, and activities.[4,5,6]

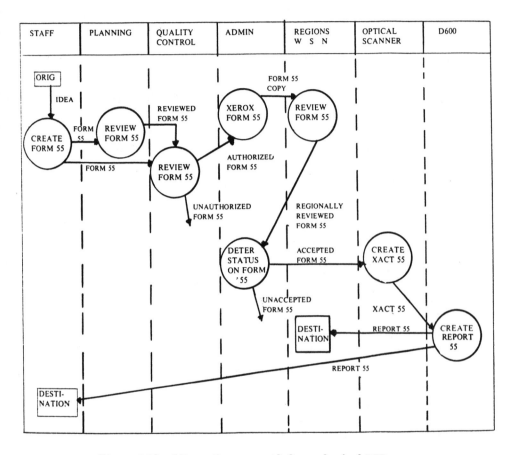

Figure 4.15. Alternative approach for a physical DFD.

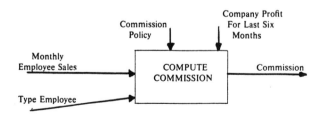

Figure 4.16. Activity Diagram.

Shown in Fig. 4.16, the Activity Diagram identifies inputs (monthly employee sales and employee type), converted by an activity (compute commission) to become an output (commission). The diagram also shows the control data (commission policy and company profit for the last six months) by arrows entering the top of the box. The control data behaves as resident information required by the activity, much like a data base does, and determines how the activity is to be performed; it is not modified by the activity.

4.7 Guidelines for drawing a data flow diagram

The data flow diagram of a program or system may involve multiple levels of diagrams, each with different amounts of detail. While we have established some guidelines for determining completion of the data flow diagraming process, we have not yet established where or how to start the process. Although analysts and designers usually know the basic inputs and outputs of the system, they may not know (and, at this time, probably should not care) whether these inputs and outputs are fixed or variable in length, are blocked or unblocked, or are on tape, card, or disk.

The data flow diagram links the input data streams to the output data streams. Analysts and designers frequently work their way from the inputs, through successive transforms, to the outputs. Working from the outputs to the inputs also is a viable approach, especially applicable when the user knows the required outputs but does not know the inputs, or when the analyst has been provided with redundant or irrelevant information about the inputs.

A third alternative is to think initially of the entire system as one black box, with all logical inputs and outputs shown, and to refine this simple diagram by decomposing it into several transforms. This approach is repeated until the desired level of detail is achieved. Figure 4.17 shows the first two levels of such a data flow diagram for an update system, developed by using this middle-out approach.

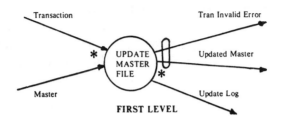

FIRST LEVEL

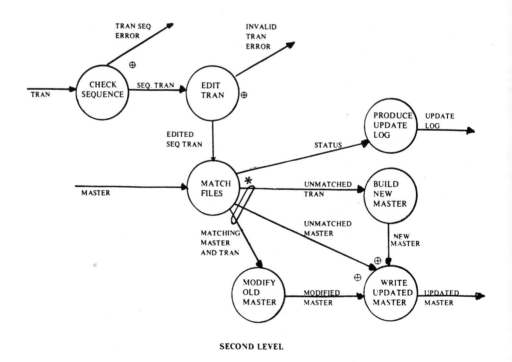

SECOND LEVEL

Figure 4.17. First two levels of a DFD showing the middle-out approach.

Whatever the chosen approach, the analyst or designer should carefully label all data streams in a data flow diagram. Without proper data-stream labeling, the subtle differences between a transaction, a sequenced transaction, and an edited, sequenced transaction easily can be missed by the document reviewer. Similarly, transforms should be described carefully whenever possible, with a specific transitive verb and nonplural object. To illuminate the point, note the difference between EDIT DELETE TRANSACTION and HANDLE DELETE TRANSACTION PROCESSING. Does the HANDLE DELETE TRANSACTION PROCESSING module access the record, edit it, sequence check it, or delete it? We just don't know, because verbs like *handle, process, prepare, do,* and *manage* usually are too imprecise.

When drawing a data flow diagram, analysts and designers sometimes find themselves thinking of decisions, control loops, error processing, initialization, and termination. These procedural aspects of the problem normally would be found in a flowchart of a system. In data flow diagrams, initialization and termination are ignored, because we are concerned with modeling the steady-state behavior of the system. The requirement for error processing is noted, but is normally not depicted in detail. While we should identify any function that filters valid data from invalid data, details for processing invalid data normally are omitted in high-level data flow diagrams. Control loops and decisions certainly exist in real computer systems, but they should not be present in data flow diagrams. Rather than thinking in terms of going back to get another record, think of the data flow diagram as showing only the processing for record 100 of a one-million-record file. Control loops and associated control logic are procedural aspects of the problem. A data flow diagram deals only with the functional aspects of the problem and the associated transformations of data.

In summary, the following are guidelines to analysts and designers for drawing data flow diagrams:

1. Identify all inputs and outputs.

2. Work your way from inputs to outputs, outputs to inputs, or from the middle out to the physical input and output origins.

3. Label all data flows carefully and descriptively.

4. Label all transforms by means of a specific transitive verb and nonplural object.

5. Clarify the association of data streams to a transform in detailed data flow diagrams by clearly indicating the appropriate logical AND and OR connectors.

6. Ignore initialization and termination.

7. Omit the details of error paths, especially in generalized levels of a data flow diagram.

8. Do not attempt to show control logic such as control loops and associated decision-making.

9. If in doubt as to the proper level of detail, assume that a further detailed breakdown of the problem is required.

In the creation of multilevel data flow diagrams:

1. Number each process within the overview DFD.

2   Identify any process within the overview DFD (the parent diagram) that requires a more detailed breakdown of function.

3. Draw a second-level DFD (the child diagram) for each process within the parent diagram that requires functional refinement, numbering each to associate the child diagram with the appropriate parent function.

4. Make sure inputs and outputs are matched between parent and associated child diagrams, except for error paths that may be present on the child but absent from the parent.

5. Repeat the above procedures for every process in the DFD, until each process has been defined in sufficient detail or in terms of its most elementary inputs and outputs.

To apply these guidelines, consider the problem of making Waldorf salad. We begin by identifying all of the inputs and outputs (subject to our tastes), and proceed as shown in Fig. 4.18:

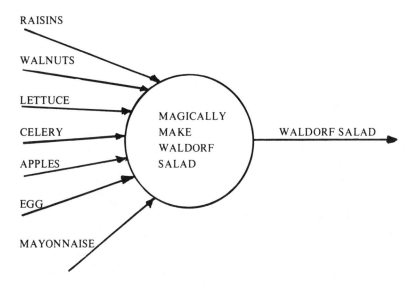

Figure 4.18. An overview DFD for making Waldorf salad.

Our overview DFD shows us inputs and output, but does not identify the detailed steps in magically making Waldorf salad. Using this one-bubble DFD as specifications, ten different chefs might develop ten distinctly different salads. Some, for example, might have raw eggs with pieces of egg shells. Other salads might have apple cores or walnut shells. The point is that if we expect people to be able to eat this salad, we had better define the detailed processes more precisely. The next functional refinement of the Waldorf salad problem could be depicted as shown in Fig. 4.19.

Our example requires even further functional decomposition to assure that ten chefs would develop essentially the same Waldorf salad; at this point, however, we can see that each of the detailed processes tends to be manageably small and functionally independent. Other details, such as how long the eggs boil, should be addressed in the detailed process specification of the BOIL EGG function. Note that the overview diagram and the second-level diagram have the same major inputs and outputs. The second-level diagram, however, identifies the processes in greater detail, includes error paths not shown on the parent diagram (rotten apples and eggs), and data bases required by the detailed processes (spices).

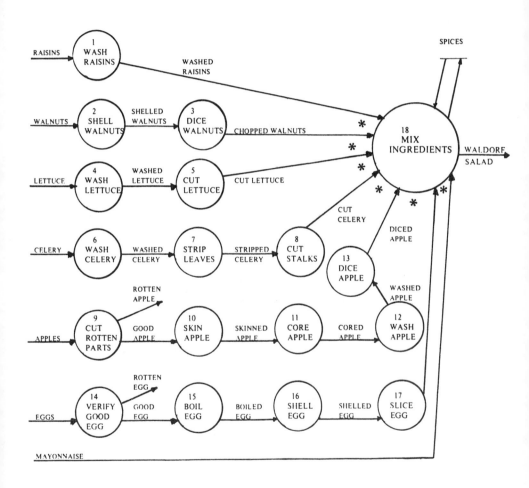

Figure 4.19. Second-level DFD for making Waldorf salad.

Review Exercise

1. Draw a physical data flow diagram to depict the inputs, outputs, and processes involved in making an omelette, baking a cake, or building, say, a sun deck.

2. Derive a logical data flow diagram from the physical data flow diagram.

3. Envision a change or changes you might make to your set of procedures.

4. Draw a logical data flow diagram that shows the new proposed system, that is, the amended instructions and associated procedures.

5. For additional practice in deriving a DFD, turn to the Appendix for the Porterhouse Shrimp Corporation Exercise.

References

1. D. Martin and G. Estrin, "Models of Computations and Systems — Evaluation of Vertex Probabilities in Graph Models of Computations," *Journal of the ACM,* Vol. 14, No. 2 (April 1967), pp. 281-99.

2. E. Yourdon and L.L. Constantine, *Structured Design: Fundamentals of a Discipline of Computer Program and Systems Design,* 2nd ed. (New York: YOURDON Press, 1978).

3. D. Teichroew and Z. Gackowski, "Documentation of Information Processing Systems: For Their Development, Operation and Maintenance," ISDOS Working Paper 169 (University of Michigan, October 1976).

4. D.T. Ross and J.W. Brackett, "An Approach to Structured Analysis," *Computer Decisions,* Vol. 8, No. 9 (September 1976), pp. 40-44.

5. D.T. Ross and K.E. Schoman, "Structured Analysis for Requirements Definition," *IEEE Transactions on Software Engineering,* Vol. SE-3, No. 1 (January 1977), pp. 6-15.

6. SofTech Inc., "Structured Analysis for Requirements Definition," 9031-9.1 (Waltham, Mass.: SofTech Inc., 1976).

5 Data Structure Diagrams

Ultimately, the users are not concerned with how the processing works, but merely that the processing works. They want assurance that transactions can be processed, and that the data base can be updated correctly. They care about the inquiries that they can make and the information that they can get in response to those inquiries.

5.1 Perspective

Who among us has not had the experience of fumbling through pockets or a handbag looking for a particular key or some important piece of paper? Typically, we find it after a few trial-and-error efforts. But what if our pockets were filled with fifty sets of keys or one hundred scraps of paper? Sooner or later, after having wasted considerable time, we probably would realize the need to organize our property more efficiently.

So long as we need to access only one item from a very limited number of items, the organization of data is not particularly important. As the potential amount of data to be accessed increases, however, we want to minimize the time spent to obtain the required information. For example, we would not want to spend hours in a library trying to find a book, or more than a minute or two trying to locate someone's telephone number. These types of informational requirements historically have been met by the creation of sequential (alphabetic) paper files. Public libraries have card index files that alphabetically arrange titles of books by author. Telephone companies publish the white pages that list telephone numbers and addresses by name. These types of sequential paper files are useful for accessing data in a limited number of ways.

Look at what happens, however, as soon as these services are provided. The users realize the benefits and imagine other ways to use the data. For example, a researcher in a library might say, "I'm not interested in the books of J.J. Jones specifically, but rather in the subject of systems analysis. Can you point me to a list of books on that subject?" Or, we may be hungry as wolves on our first night in a new town and may not know the names of any local eating establishments. What we want is a pizza delivered, not a lengthy search through the white pages of a telephone directory.

Fortunately, both public libraries and telephone companies have anticipated these consumer needs and provide the appropriate services. Libraries have an index listing titles of books by subject. Telephone companies have the yellow pages. The variation in the way the data could be used has resulted in the data being reorganized and duplicated. But, how many times will these organizations reorganize the data? A historical researcher may be interested in all books published between 1620 and 1640. The standard organization of data in a library may not provide easy access to that information. Should a library again duplicate and reorganize the data to list titles for each author by date of publication? Should telephone companies create pink pages to list telephone numbers and establishments by street address?

The point is that organizations now are recognizing the value of manipulating massive amounts of data in many ways. Access to data now is required beyond traditional project or departmental boundaries. For example, project management may want to access project data to determine how actual progress compares to estimated progress. Another project's management may want to access the same data to obtain historical perspectives, perhaps to identify problems to avoid in future projects. A personnel department might want to view the same data to study personnel utilization trends. A marketing department might want to do marketing analysis of the data in order to establish future plans.

As the volume of data and the number of ways in which data are used increase, speed of data retrieval and the costs involved in data duplication and reorganization become critical. Many organizations have replaced sequential paper files with sequential tape files and disk files, which can be accessed either sequentially or randomly to cut down on the time needed to access data. These organizations also may have created data base facilities that provide hierarchical records, which are accessible by more than one key to minimize both data duplication and reorganization costs.

5.2 Data base

Before further discussing data structure requirements, we should clarify what is meant by the term *data base*. From the point of view of its users, a data base is a reservoir of information that supplies users with the data from which they make decisions.[1] In its broadest sense, a data base is a collection of data used for many applications within an organizational environment. From a data processing point of view, a data base is a collection of data stored by computer hardware and manipulated by programs.[2] A data base could refer to data retained from transaction to transaction, such as temporary tables or temporary data sets. It also could refer to data retained from program load to program load, such as a data bank or permanent data set. We should keep in mind, as well, that a data base need not necessarily involve an automated process or high-speed access. We can think of a telephone book or a card index file of customer names and addresses as a data base. Nowadays, however, a data base usually is thought of in terms of an integrated body of data utilized in an automated data processing environment, thus the term *integrated data base*. An integrated data base is a "collection of interrelated data stored together with controlled redundancy . . . to serve one or more applications . . . independent of programs which use the data. . . ."*

---

*James Martin, *Principles of Data Base Management* (Englewood Cliffs, N.J.: Prentice-Hall, 1976), p. 4.

5.3 Defining data base requirements

Before we discuss data base requirements, let us look at Fig. 5.1, which shows how an integrated data base typically is used: to support regularly scheduled work and to answer unscheduled inquiries (ON-DEMAND), made by the users of the system. Inquiries can be made in a PREDEFINED format or by using an interrogation language such as MARK IV. The problem is that the organization of the data base may not provide quick and efficient retrieval of data. Sorting the data base information into the required format may take too long and cost too much. As a result, the analyst's main concern in identifying data base organization requirements is to assure that all ON-DEMAND inquiries in a PREDEFINED format, which require IMMEDIATE responses, have been clearly defined and can be met within the specified time and cost constraints.

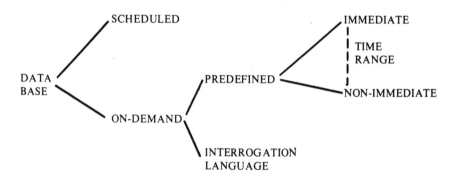

Figure 5.1. Data base usage.

Data base technology is a specialized field within the already specialized field of automated systems development. Frequently, users have considerable problems keeping up with the developments within their respective business areas. They cannot be expected to understand in addition the physical structure of a data base. Moreover, they do not need to know and, most often, do not want to know about that physical structure. They want to view the data base as a black box within the data processing environment, as shown on the next page in Fig. 5.2:

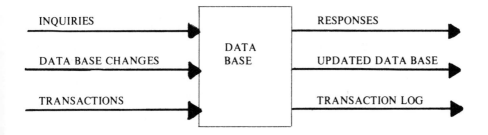

Figure 5.2. Data base black box.

Ultimately, the users are not concerned with how the processing works, but merely that the processing works. They want assurance that transactions can be processed, and that the data base can be updated correctly. They care about the inquiries that they can make and the information that they can get in response to those inquiries. These requirements, especially those relating to inquiries and responses, will have a strong influence on the data base design. Unfortunately, there are four basic problems in defining data base requirements:

1. Users do not realize the level of precision required to develop well-stated objectives in the form of specific inquiries and responses.

2. Users have difficulty detaching themselves from their current way of doing business to imagine the inquiries that they might be able to make.

3. Users generally are not aware of the time, space, and financial trade-offs associated with the retrieval of data.

4. Different users want different information from the system, or the same information presented in different ways.

Systems analysts frequently are no more qualified than users to develop physical data base systems. Nowadays, there are literally hundreds of data base systems (IMS, CICS, TOTAL, ROSCOE, among others) with unique characteristics to serve specific needs. Furthermore, data base technology is moving so far so fast that generalists, such as systems analysts, cannot be expected to keep up with this dynamic subject.

While systems analysts frequently are responsible for developing the physical design of applications systems, they hardly ever are responsible for the physical design of data base systems. That job is the domain of the data base designer. While the systems analyst sees data base requirements from an applications viewpoint, the data base designer must consider all of the applications served by the data base.

Systems analysts, however, can be effective in defining user requirements, in developing logical specifications, and in verifying that the physical data base design meets the objectives of the users (see Fig. 5.3 on the next page). Our main focus should not be to develop the physical data base design, but rather to apply our analytic skills and data processing experience to develop a precise information requirements specification. We can bring technical awareness and creativity to the problem to identify the possibilities for information retrieval and the associated time, space, and financial trade-offs. We can use analytic tools, such as questionnaires, to establish the relative importance of inquiries as viewed by the various users of the system.

Once this is done, we should have sufficient perspective to develop a consensus of user objectives; we also should communicate them in the form of a graphic, logical specification supported by precise requirement definitions that relate to the characteristics of inquiries (response time, volume, peaks, valleys, security, and so on) and to the constraints of development (equipment, time, costs, and benefits).

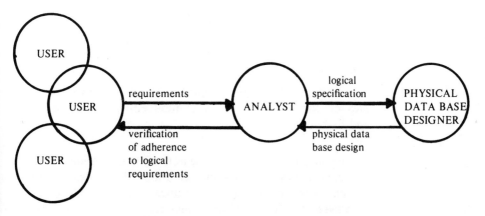

Figure 5.3. Data base design roles.

5.4 Data structure diagrams

The data structure diagram is a graphic tool to be used by the analyst to describe the users' logical data structure requirements. Introduced in Chapter 3, the DSD is intended to be used both to verify informational requirements with the users and to communicate them to a physical data base designer.

Historically, systems analysts have thought of data as existing in the form of sequential files, records, and fields. In describing data base characteristics, it is useful to refine and expand our definitions to identify unique aspects of data base data structures. These definitions provide the building blocks for the DSD. Some data base theorists have replaced file, record, and field terminology with the terms *universe, individual,* and *attribute,* respectively.[3] Others have described data base characteristics in terms of objects, properties, and object relationships. The building blocks in a data structure diagram are known as *entities, records, attributes, key attributes, attribute pointers,* and *logical pointers.* These terms are defined as follows:

ENTITY: a general class of items about which information is recorded. An example is a department entity containing relevant data for each department in the environment.

RECORD: a specific set of items within a general class of information. An example is the set of information pertaining to the Accounting Department, one of many departments defined in the department entity.

ATTRIBUTE: a property possessed by a record or an entity. Two examples of attributes of the department entity are its name and its budget.

KEY ATTRIBUTE:

an attribute that is used as a search argument to identify a particular logical record within an entity. Examples of key attributes are the employee identification number (for the employee entity) and the department name (for the department entity).

ATTRIBUTE POINTER:

an attribute that is used as a pointer to a particular set of information (logical record) in another entity, and that exists in more than one entity. An attribute pointer in one entity will be a key attribute in another entity. As an example, the employee names in the department entity are used as attribute pointers to the employee name, which is a key attribute in the employee entity.

LOGICAL POINTER:

a pointer (graphically represented by an arrow) that identifies the relationship between entities. It indicates the ability to gain immediate access to the information in one entity (for example, the employee entity) by initially defining a key attribute in another entity (the department name in the department entity).

Figure 5.4 depicts the terms described above, and shows a data structure diagram for a logical data base structure. The diagram indicates two data entities, each represented by a rectangular box — one entity comprising departments and one comprising employees. Each record within the DEPARTMENT entity is uniquely identifiable by name of department (DEPT-NAME), the key attribute. Each record within the EMPLOYEE entity is uniquely identifiable by either of two key attributes: the name of the employee (EMPL-NAME) or the employee identification number (EMPL-ID-NO). It is a useful graphic convention to separate the entity name from the key attributes by a solid line. Attributes that are neither key attributes nor attribute pointers follow the key attributes. Attributes are separated from key attributes by a dotted line. Attribute pointers are shown at the bottom of the box depicting an entity and are separated from other attributes by a dotted line. The arrow between the DEPARTMENT and EMPLOYEE entities indicates a logical relationship between the entities: By identifying DEPT-NAME, the user will be able to gain immediate access to information in the EMPLOYEE entity. It further indicates a one-way street in that the user will not be able to gain immediate access to the DEPARTMENT entity by specifying an employee name or an employee identification number. In Fig. 5.4, note that the (S) after EMPL-NAME in the DEPARTMENT entity indicates the possibility of more than one employee name.

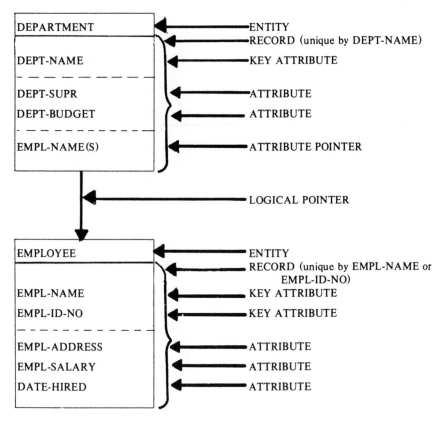

Figure 5.4. DSD depicting a logical design to access departmental records and personnel records by department name, by employee identification number, or by employee name.

Let us assume that as a result of additional meetings with users, systems analysts have identified other requirements. Users want immediate access to project information and to employee records by specifying a project identification. Figure 5.5 shows how those requirements might be pictured graphically in a data structure diagram.

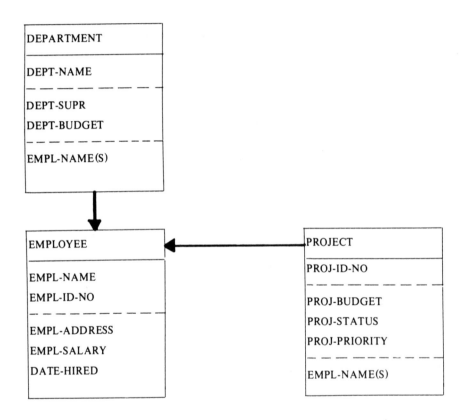

Figure 5.5. DSD showing expansion of requirements in Fig. 5.4.

One can imagine that additional meetings with users or that changing patterns in the business might result in the identification of new information requirements. For example, analysts might determine that the users also need immediate responses to such questions as those listed on the following page:

- What projects is John Doe assigned to?

- Who was hired the same day as John Doe?

- Who is John Doe's department supervisor?

Figure 5.6, an expansion of the logical data base design already depicted in Fig. 5.5, shows the logical capability to answer these questions immediately.

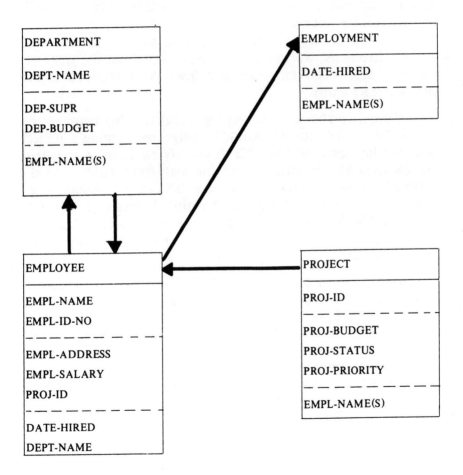

Figure 5.6. DSD showing expansion of requirements in Fig. 5.5.

Look again at the questions listed on the previous page, and at Figs. 5.5 and 5.6. Adding the project identification (PROJ-ID-NO) as an attribute of the EMPLOYEE entity will enable us to answer the first question. By using DATE-HIRED in the EMPLOYEE entity as an attribute pointer and by creating a separate EMPLOYMENT entity with DATE-HIRED as a key attribute and EMPL-NAME(S) as an attribute, we can identify all other employees hired on the same date. Note that the logical relationship between the EMPLOYEE entity and the EMPLOYMENT entity is shown by a logical pointer. By creating DEPT-NAME as an attribute pointer in the EMPLOYEE entity, we will be able to create a two-way bridge between the EMPLOYEE and DEPARTMENT entities. Because the logical pointers between these two entities go in both directions, we will be able to go both from the DEPARTMENT to the EMPLOYEE entity and from the EMPLOYEE to the DEPARTMENT entity.

Figure 5.6 shows a one-way logical relationship between the EMPLOYEE entity and the PROJECT entity. By defining a project identification, users will be able to identify employee names that can be used as key attributes to the EMPLOYEE entity. As diagramed, however, Fig. 5.6 does not indicate the ability to go from the EMPLOYEE to the PROJECT entity. Before reading further, study Fig. 5.6 to identify the change that would be required to show a logical relationship from the EMPLOYEE entity to the PROJECT entity.

Figure 5.7 depicts the revision that would be required. Note that the PROJ-ID, identified as an attribute of the EMPLOYEE entity in Fig. 5.6, now has been defined as an attribute pointer in Fig. 5.7. Since the PROJ-ID already had been defined as a key attribute of the PROJECT entity in Fig. 5.6, no changes are required within the PROJECT entity. The only differences between Figs. 5.6 and 5.7 are the definition of the PROJ-ID as an attribute pointer and the resulting logical relationship as shown by the logical pointer from the EMPLOYEE to the PROJECT entities.

You may be wondering about the apparent lack of precision related to sources and destinations of logical pointers. Students in structured analysis seminars and even reviewers of drafts of this book have asked, Do arrows (logical pointers) go from the bottom of a box (entity), from the top of a box, or do they emanate directly from an attribute pointer? Do these logical pointers go to the bottom of a box, to the top of a box, or to the associated key attribute?

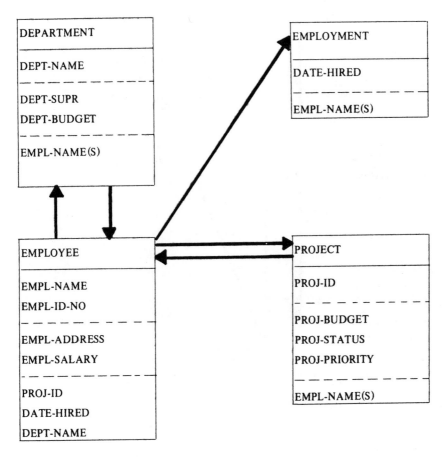

Figure 5.7. DSD showing revision to Fig. 5.6.

Logically, the logical pointers go from the attribute pointer to the associated key attribute in another entity. Physically, however, logical pointers in a data structure diagram should be drawn in such a way as to minimize the crossing of lines and to maximize the clarity of the diagram. Keeping in mind that Fig. 5.7 pictures a relatively simple set of data structure requirements, note how the meaning of the diagram might be obscured when logical pointers are drawn from the attribute pointer to the key attribute, as shown in Fig. 5.8. While the logical relationships in Fig. 5.8 could be drawn without crossing any lines, this graphic approach will result in unnecessarily complicated diagrams for even more complex, real-life data base systems.

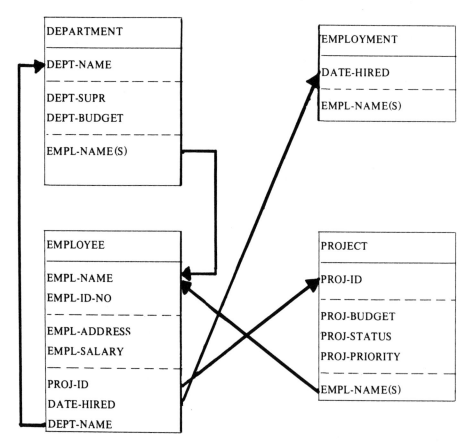

Figure 5.8. DSD for logical pointers from attribute pointer to key attribute.

Although the logical data structure diagram may appear to organize data into fixed physical relationships, this is, in fact, not the case. For example, although we have duplicated EMPL-NAME in each entity in Fig. 5.7, we are neither suggesting nor dictating that this duplication should exist in the physical data base. We do not see data base requirements from an overall organizational perspective, but only in terms of our specific application. Neither do we understand the trade-offs between speed of access and costs of data duplication. These considerations ultimately must be addressed by the physical data base designer.

5.5 Guidelines to develop a logical data base specification

The derivation of a logical data base design is not a straightforward process. As already stated, users frequently have difficulty envisioning their data base requirements. In addition, different users want from the system different information in different time frames. Analysts can aid in the process of logical data base definition by adhering to the following procedures:

1. Meet with all relevant users to gather information on their data base informational requirements and objectives.

2. Develop a data structure diagram from the initial information provided by users.

3. Develop a list of inquiries that could be answered immediately, given the DSD.

4. Develop a list of inquiries that could not be answerable immediately, given the DSD. (It is essential to make clear not only what the system will do, but also what the system cannot do.)

5. Draw up a questionnaire, as shown in Fig. 5.9, for all inquiries that both might and might not be answerable immediately.

6. Discuss each of the inquiries on the questionnaire with each important member of the user community to identify user ratings.

7. Analyze and consolidate the ratings.

8. Revise the DSD according to the rating information gathered.

9. Return to the users to verify the completeness of the logical data requirements as shown in the revised data structure diagram.

10. Identify the time, space, and financial trade-offs associated with the physical implementation of the logical data base design.

11. Get concurrence from the users on the consensus of logical data base objectives.

12. Develop a logical data base specification for a physical data base designer that includes:
    - lists of inquiries to which the system must respond immediately
    - other anticipated inquiries
    - a logical data structure diagram
    - a logical definition of all data required to answer inquiries
    - volume of inquiry requirements
    - special characteristics of volume (peaks)
    - security and privacy requirements
    - resource constraints
    - cost constraints
    - time constraints

INQUIRY	SYSTEM INPUT	SYSTEM OUTPUT	RELATIVE VALUE OF INFO instantly ǀ tomorrow	
			(scale of 1 — 10, 10 most important)	
1. Project info	PROJECT-ID	PROJ-BGT PROJ-STATUS PROJ-PRIORITY EMPL-NAME(S)	9	2
2. Department info	DEPT-NAME	DEPT-SUPR DEPT-BGT EMPL-NAME(S)	8	3
3. etc.				

Figure 5.9. Questionnaire for anticipated data base requests.

Review Exercise

1. Identify a situation in which a system utilizes a data base in each of the following ways:
    a) as data held from transaction to transaction
    b) as data held throughout the running of a program
    c) as an integrated data base

2. Identify the role of the systems analyst in the creation of data base systems.

3. Assume that as a result of preliminary meetings and structured interviews with all users of a banking system, you have identified a list of typical questions that the data base of the system must be able to handle with immediate responses. Your task is to draw a DSD that will describe the logical data structure requirements of the system.

    Note: In doing this problem, assume the following:

    - *An affiliated group is synonomous to a conglomerate.*

    - *Only the member corporations, not the affiliated group, have an account with the particular bank.*

    - *Each member corporation has only one account with the bank.*

    While the questions below are about specific companies and accounts, they represent typical questions about all companies and accounts associated with the bank.

    Typical questions requiring immediate response are:

    a) Who are the corporate members of the ABCD Industries' affiliated group? (From your point of view, the question is, Given the name of any affiliated group, what are the member corporations?)

b) What affiliated group is associated with the WWW Corp.?

c) What are the address and telephone number of the affiliated group associated with corporate member WWW Corp.?

d) Corporate account number 12345 has just bounced five checks. Who are the officers and how do we contact them?

e) What is the current available balance for WWW Corp.?

References

1. B. Sundgren, *Theory of Data Bases* (New York: Petrocelli/Charter, 1975).

2. G. Wiederhold, *Data Base Design* (New York: McGraw-Hill, 1977).

3. I. Flores, *Data Structure and Management* (Englewood Cliffs, N.J.: Prentice-Hall, 1970).

6 Tools to Express Logic

Specifications have been notoriously incomplete, ambiguous, and contradictory.

6.1 Perspective

Systems analysts frequently must communicate the logical requirements of a system both to the users for verification and to the designers and programmers for implementation. Sometimes, these logical requirements do not necessitate detailed explanations. When a date field must be verified to be a valid date or when an account number field must be verified to be numeric, no further details of the required functions need to be provided. Programmers know, for example, that thirty days hath September, April, June, and November and that a date validation routine has to account for leap year. It even might be counterproductive for the systems analyst to outline the rules for date validation to either the users or the programmers. Users are interested only that the date be validated, not in how it is done. Programmers have to know simply that a date verification is required; they might interpret any further instructions from the systems analyst as a violation of their domain, an infringement upon their programming creativity.

Most logical requirements, however, do require detailed explanation. While programmers already know the rules associated with dates, what do they know about the nature of the business being automated? What do they know, for instance, about the credit, commission, and pricing policies in a particular business? While the same rules apply to validate a date from one system to another, policies change from system to system and from business to business. These unique systems' processing requirements must be defined in detail.

Normally, detailed processing requirements are written by the systems analyst in the form of program or module narratives, which tend to be the source of considerable misinterpretation. The more complex the processing rules are, the more likely these narratives will be ambiguous and perhaps even incomplete. One way to assure that these processing requirements are both clear and complete is by applying logical tools to aid in their definition, and thereby provide well-defined conditions and actions in the form of a graphic that can be easily comprehended.

Three types of logical tools are presented in this chapter — structured English, decision tree structures, and decision tables. These tools should be used to support, if not replace, narrative program and module specifications. Typically, the specifications expressed by these tools will be reviewed with relevant users to identify potential omissions, ambiguities, and contradictions in a policy. These specifications also will be given to programmers to support narrative program specifications and to permit easy, methodical coding.

The proper use of these tools depends upon the nature of the problem, the expertise of the people reviewing the solutions, and the particular software available at an installation. For example, structured English may be the most "long-winded" of the three tools, but it also is closest to English and to COBOL. Decision tree structures clearly indicate the processing requirements and, generally, necessitate the least amount of training of both the tool user and the document reviewer. Decision tables, while perhaps a bit more difficult to create and comprehend than the other logical tools, can be used directly as input to create program logic with the aid of software preprocessors. [1]

Choosing a specific tool to use, however, is not as important as the fact that detailed logical processing requirements should be defined by the analyst. Too often, programmers spend their time not programming, but rather debugging the ambiguous and incomplete specifications received from the analyst. Contrary to popular belief, the programming task should not involve expanding or revising program specifications by the use of flowcharts, decision tables, or narrative descriptions.[2] These should be provided by the analyst so that the programming task is confined to coding and debugging the system.

6.2 Structured English

Structured English, also known as pseudocode, computer Esperanto, or Program Design Language, is an English language adaptation used to describe logical program requirements. It is an extension of a basic assumption of structured programming. Just as any program can be written with only three basic structures, the logic of any process can be expressed by any of these same three structures: imperative statements, decision statements, and repetition statements. Using these statements, let us now identify how to create a structured English specification and the guidelines to assure that specifications are as clear and as error-free as possible.

Imperative statements tell us to do something, such as, Process all good transactions. Unfortunately, what to do may not be clear to the programmer, because he may not know what the processing involves. Even if he did, he might have a question about what is meant by a "good" transaction.

To correct these potential problems, the guidelines of structured English suggest that imperative statements be written according to the following rules:

- The statement should be concise. (Avoid long, rambling sentences with connectors such as *unless, but, however,* and *except.)*

- The statement should contain a verb that clearly describes the required function. (Avoid verbs like *do, handle, process,* and *control.)*

- The object of the statement should be stated explicitly. (Specify what is to be edited, printed, or computed.)

- All nouns should be documented in a data dictionary for the system.

- The use of adjectives should be limited to words that previously have been defined or that are self-explanatory.

- The use of adjectives, adverbs, and verbs expressing relativity should be minimized. (Avoid words like *increase, reduce, improve, faster, more,* and *good.)*

Figure 6.1 features imperative statements in a structured English description for an UPDATE-MASTER module. Note that all statements are concise and clear. All of the verbs clearly define the processing requirements. All of the nouns are descriptive names of data fields, which should be documented in the data dictionary for the system. Adjectives and adverbs that might make the processing requirements subject to interpretation are notably absent.

UPDATE-MASTER

      Move TRAN-ADDRESS to MAST-ADDRESS

      Move TRAN-CYCLE to MAST-CYCLE

      Move TRAN-LAST-DATE to MAST-LAST-DATE

      Add TRAN-NEW-CHARGES to MAST-OWED

      Subtract TRAN-PAYMENT from MAST-OWED

END-OF-UPDATE-MASTER

Figure 6.1. Structured English imperative statements.

Figure 6.1 would provide the programmer responsible for the UPDATE-MASTER module with a precise module specification; however, it is too heavily oriented toward data processing to be of much value to the user. Keep in mind that the usefulness of the document relates both to the nature of the problem and to the interests of the potential reviewers.

Decision statements in structured English are analogous to the IF-THEN-ELSE construct of structured programming. The binary decision-making process is expressed by the IF-ELSE construct in structured English. The N-way decision-making process is expressed either by strings of IF-ELSE constructs or by the CASE construct (see Fig. 6.2a, b, and c).

```
IF CHOCOLATE-ICECREAM
            give me a scoop
ELSE
            give me the check
```

a. Binary decision-making.

```
IF CHOCOLATE-ICECREAM
            give me a scoop
ELSE (no CHOCOLATE-ICECREAM)
IF VANILLA-ICECREAM
            give me a scoop
ELSE
            give me the check
```

b. N-way decision-making (IF-ELSE).

There are 3 CASES

```
CASE 1   CHOCOLATE-ICECREAM
            give me a scoop

CASE 2   VANILLA-ICECREAM
            give me a scoop

CASE 3   None of the above
            give me the check
```

c. N-way decision-making (CASE).

Figure 6.2. Binary and N-way decision-making process constructs.

To see how decision statements in structured English are derived, let us take a look at the specification narrative that follows in Fig. 6.3.

Whenever the insured is greater than 25 years old and has had more than one claim in the past year, add $200 to the amount due. In the case of the insured having had more than one claim in the past year and being not older than 25 years, then add $400 to the amount due. When the insured has not had more than one claim in the past year, add $50 to the amount due, unless the insured is older than 25, in which case add $25 to the amount due.

Figure 6.3. Specification narrative of insurance amount-due problem.

Under what conditions will $50 be added to the amount due? Working with the given information, we normally would backtrack through the specification narrative. Unfortunately, this approach is time-consuming, frustrating, and prone to error. The problem, however, is not in the backtracking approach itself, but rather that the specification narrative is difficult to follow.

By using structured English, we can improve the readability of our specifications. Figure 6.4, a rewrite of Fig. 6.3 in structured English, shows the use of both indention and parenthetical expressions to enhance readability. Indention helps to isolate each set of unique conditions and actions. The use of parenthetical expressions after the ELSE in the IF-ELSE construct results in an explicit statement of the conditions associated with the ELSE part of the logic. From the structured English notation in Fig. 6.4, it is now relatively easy to identify the conditions under which $50 will be added to the amount due. With the help of indention and parenthetical expressions, we can see that $50 will be added to the amount due only when the insured party has not had more than one claim in the past year *and* when the insured is not older than 25. Note that readability is enhanced by connecting an IF with its associated ELSE by means of a dotted line. Note also that the dotted-line notation insures that for every IF clause, there is a corresponding ELSE clause.

IF CLAIMS-PAST-YEAR exceeds 1 and AGE exceeds 25
      add $200 to AMOUNT-DUE

ELSE

IF CLAIMS-PAST-YEAR exceeds 1 and AGE does not exceed 25
      add $400 to AMOUNT-DUE

ELSE

IF CLAIMS-PAST-YEAR does not exceed 1 and AGE exceeds 25
      add $25 to AMOUNT-DUE

ELSE (CLAIMS-PAST-YEAR does not exceed 1 and AGE does not exceed 25)
      add $50 to AMOUNT-DUE

Figure 6.4. Structured English of insurance amount-due problem.

Figure 6.4 shows the use of strings of IF-ELSE constructs. The same logical problem can be depicted either by the use of the CASE construct or by nested IF statements, as shown in Fig. 6.5. A nested IF statement is an IF-ELSE construct that exists *within* the boundaries of at least one other IF-ELSE construct. Strictly following this definition, the reader may have difficulty in finding both nested IF statements in Fig. 6.5: The construction of "CLAIMS-PAST-YEAR does not exceed 1" as a parenthetical expression in Fig. 6.5, in fact, hides an IF-ELSE construct.

IF CLAIMS-PAST-YEAR exceeds 1
   IF AGE exceeds 25
        add $200 to AMOUNT-DUE
   ELSE (AGE does not exceed 25)
        add $400 to AMOUNT-DUE
ELSE (CLAIMS-PAST-YEAR does not exceed 1)
   IF AGE exceeds 25
        add $25 to AMOUNT-DUE
   ELSE (AGE does not exceed 25)
        add $50 to AMOUNT-DUE

Figure 6.5. Nested IFs of insurance amount-due problem.

Figure 6.6, logically equivalent to Figs. 6.4 and 6.5, reveals both of the nested IF statements in the insurance amount-due logic problem. While the logic shown in Figs. 6.5 and 6.6 is not particularly difficult to follow, studies have indicated that two levels of nested IF statements are the most that people can be expected to comprehend comfortably.[3] The point about nested IF statements is not that they should be abolished. Instead, they should be recognized as useful, but potentially complex constructs, especially when logic is depicted with multiple levels of nested IFs. A systems analyst would do well to read aloud the logic that has been represented by nested IF statements. If the analyst has difficulty in following the logic, it should be rewritten into some other format. As an example, excessively complicated nested logic can be reformatted by combining conditions in the form of complex statements to remove the nesting (IF A and B and C and D). Keep in mind that the ultimate question is not whether the systems analyst can follow the logic, but whether all other relevant parties can follow it.

```
IF CLAIMS-PAST-YEAR exceeds 1
    IF AGE exceeds 25
            add $200 to AMOUNT-DUE
    ELSE (AGE does not exceed 25)
            add $400 to AMOUNT-DUE
ELSE
IF CLAIMS-PAST-DUE does not exceed 1
    IF AGE exceeds 25
            add $25 to AMOUNT-DUE
    ELSE (AGE does not exceed 25)
            add $50 to AMOUNT-DUE
ELSE (end-of-action)
```

Figure 6.6. Nested IFs revealed of insurance amount-due problem.

Repetition statements indicate the conditions under which a set of activities will be repeated and the conditions under which the repetition will be terminated. In structured English, such iterations can be specified either by use of the DO-WHILE construct or the REPEAT-UNTIL construct. These constructs are equivalent, except that the REPEAT-UNTIL construct implies that the loop body logic must be executed at least once before the loop is exited, while the DO-WHILE construct tests a condition before the loop body is executed. This is shown in Fig. 6.7, below:

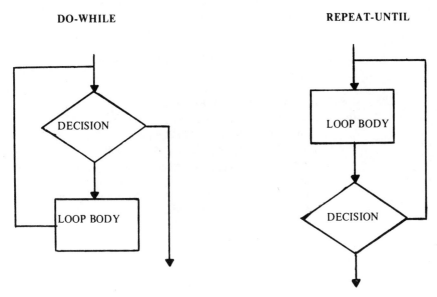

Figure 6.7. DO-WHILE and REPEAT-UNTIL constructs.

Figures 6.8 and 6.9, on the following page, show identical policies expressed in different ways using structured English with imperative, decision, and repetition statements. Figure 6.8 expresses a stock-option policy at a level and in a language to which users can relate. It should provide a permanent record of both the analyst's and the users' understanding of that stock-option policy by being documented as part of the logical specifications in the data dictionary.

```
REPEAT-UNTIL there are no more EMPLOYEE records
       IF EMPLOYEE has been with the company full-time
       |        for at least 5 years and has
       |        not been granted any stock

       |     IF EMPLOYEE earns at least $20,000 per year
       |     |     generate STOCK-OPTION-CONTRACT
       |     |     establishing 1 share per $100 annual
       |     |     salary (round down to the nearest
       |     |     hundred — example $22,650 = 226 shares)
       |
       |     ELSE (employee makes less than $20,000 per year)
       |          do not generate a STOCK-OPTION-CONTRACT
       |
       ELSE (not full-time employee for at least 5 years
             or EMPLOYEE already has been granted stock)
                  do not generate a STOCK-OPTION-CONTRACT
```

Figure 6.8. Structured English for users.

```
REPEAT-UNTIL no more EMPLOYEE-RECORDS
    |     IF EMP-ORIG-DATE precedes CUR-DATE by
    |     |     at least 5 years and EMP-STOCK equals
    |     |     0 and EMP-SALARY equals at least $20,000
    |     |        MOVE CUR-DATE to OPTION-DATE
    |     |        MOVE EMP-NAME to OPTION-NAME
    |     |        DIVIDE EMP-SALARY by 100 (ignore remainder) and
    |     |        MOVE result to OPTION-SHARES
    |     |        WRITE STOCK-OPTION-CONTRACT
    |     |        GET EMPLOYEE-RECORD
    |     |
    |     ELSE
    |     |
    |     |        GET EMPLOYEE-RECORD
    |     END-IF
    |
END-REPEAT-UNTIL
```

Figure 6.9. Structured English for programmers.

Figure 6.9 expresses the same policy in the form of detailed programming specifications. Once the users have agreed to the policy as stated in Fig. 6.8, they probably will not be interested in the level of detail shown in Fig. 6.9. (Many times, the specification for user review will be sufficiently precise so that more detailed specifications for programmers may not be required.) The level of detail shown in Fig. 6.9, however, might be required for both programmer specifications and data dictionary documentation of the system's physical implementation. Some programmers also use the pseudocode represented by Fig. 6.9 as comments in the program listing. This approach, while having the advantage of listing the specifications and the supposedly matching code side by side, also results in duplicating the program logic and in compounding maintenance-update problems. Rather than suggest a firm policy regarding the maintenance and placement of pseudocode, we prefer to stress that the level of documentation should be directly related to the job's magnitude and complexity. Under all circumstances, however, documentation must be updated to reflect properly the system's function.

The goal of structured English is to improve communication by enhancing the readability of logic documentation. Some people have suggested that AND-IF, SO, and THEN should be included in structured English notation to make our documents more like everyday English. Figure 6.10 shows how structured English would read including these notations. Their use, however, is not necessarily recommended: While the structured English narrative may be easier to read, these additional notations may partially obscure the essence of the problem.

```
IF DISTRICT equals 1
|    AND-IF DEPT equals SALES
|         |    THEN add $15,000 to BUDGET-AMOUNT
|         ELSE (DEPT not SALES)
|              SO add $20,000 to BUDGET-AMOUNT
ELSE (DISTRICT not 1)
     SO add $70,000 to BUDGET-AMOUNT
```

Figure 6.10. Structured English including AND-IF, SO, and THEN.

Those who suggest the use of AND-IF, SO, and THEN to enhance the readability of structured English also might suggest the elimination of the END-IF, END-DO-WHILE, and END-REPEAT-UNTIL constructs, claiming that these constructs are foreign to English-speaking users and, therefore, require some user training or, at least, explanation. These constructs, however, do provide a function by delimiting a block of logic. For example, looking at Fig. 6.11, we might ask ourselves, Under what circumstances will 1 be added to the CALL-COUNT?

```
IF CALL has at least two TIMING-ENTRIES
|   compute ELAPSED-TIME (ANSWER-TIME minus DISCONNECT-TIME)
ELSE (no or one TIMING-ENTRY)
IF CALL has one TIMING-ENTRY
|   move MINIMUM-DURATION to ELAPSED-TIME
ELSE (no TIMING-ENTRY)
|   create UNANSWERED-CALL
END-IF
add one to CALL-COUNT
```

Figure 6.11. Structured English including END-IF construct.

Both the existence of the END-IF construct and the lack of indention associated with the "add one to CALL-COUNT" statement identify that 1 will be added to the CALL-COUNT no matter how many TIMING-ENTRIES are associated with a CALL. If the END-IF construct were eliminated, we still would have the indention (or lack of it) to guide us to the meaning of the policy. However, we invite trouble when we rely solely on indention to express logical policies. To compensate for potential ambiguity in situations in which we do not use the END-IF (or END-DO-WHILE) construct, we could replace it with blocking notation to clarify the boundaries of our logic (see Fig. 6.12). This notation can be used to delimit blocks of logic, and, therefore, blocks of logic within blocks of logic.

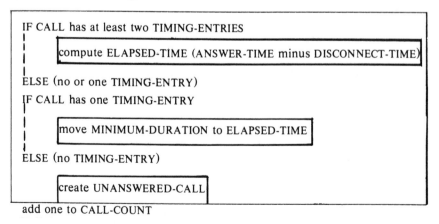

Figure 6.12. Structured English with blocking notation.

6.3 Guidelines to develop structured English

In most systems development projects, the English language or some adaptation, such as structured English, will be required to express specifications. Our job is not to create sentences featuring proper grammar, but rather to develop specifications characterized by clarity and precision. Effective communication usually results from complete, but simple, explanations. The more complex the explanations or specifications are, the more likely it is that they will be misunderstood. Keeping in mind that simple usually is equivalent to efficient, we should consider the following guidelines for structured English:

1. Avoid long, rambling statements.

2. Avoid the use of adjectives and adverbs.

3. Minimize the use of verbs expressing relativity.

4. Indent for readability.

5. Use dotted-line notation to connect an IF with its associated ELSE.

6. Use parenthetical expressions to clarify the conditions associated with the ELSE part of a statement.

7. Use parenthetical expressions to clarify the meaning of complex statements containing mixed AND and OR connectors.

8. Use either END-IF (END-DO-WHILE, END-REPEAT-UNTIL, and so on) or blocking notation to delimit a block or blocks of logic.

9. Do not not use non-negative logic wherever possible. (Isn't positive logic simpler?)

10. Avoid the use of excessive, nested logic.

6.4 Decision tables

A decision table is a graphic tool that identifies the conditions in a problem and the actions resulting from each unique set of conditions.[4,5,6] While structured English emphasizes the use of an adaptation of the English language to describe these conditions and actions, a decision table presents the problem in tabular format. Decision tables are especially applicable to problems involving many interacting decisions and possible actions. An advantage of the decision table approach is its tendency to assure that all possible sets of conditions and actions have been properly defined. As a result, using decision tables typically produces precise specifications that are prone neither to error nor to omission.

If a problem narrative totally defines the conditions and actions in the problem, the resulting decision table does not give us new information, but rather clearer information because extraneous and redundant information is deleted. For example, clauses starting with words like *but, except,* and *however* are not present. Duplicate ways of saying the same thing are compressed to form standardized terminology. Consequently, the programmer can spend more time implementing the specification and less time debugging it.

The fact is, however, that many problem narratives do not totally and properly define the conditions and actions in the problem. Specifications notoriously have been incomplete, ambiguous, and contradictory. Given this historical perspective, the analyst's use of decision tables will result in not only a clear representation of the problem, but also possibly identifying whether the problem has been thoroughly and properly defined. This frequently is not apparent from merely reading the specification. Once the specification has been broken down into a decision table format, the analyst easily can recognize the omissions, contradictions, and ambiguities in the problem statement. At this point, the analyst should be prepared to go back to the source of the problem narrative for clarification.

6.4.1 Guidelines for drawing a decision table

Now that we are familiar with the basic concepts and advantages associated with decision tables, let's concentrate on how to create a decision table. Consider the following problem:

The minimum rates for stockbrokers' commissions on stocks and warrants sold on the New York, American, and other major stock exchanges, which had been established on November 19, 1974, were abandoned on May 1, 1975, in favor of completely competitive rates.

Other stock brokerage houses substantially have reduced or revised their commission rates, and we at ABC Brokerage, Inc., must adjust our rates accordingly to keep in step with the changing business environment. Our Research and Development Department developed a plan for a commission rate schedule, to become effective July 1, 1978. Please review the following paragraph from that plan to identify any inconsistencies, ambiguities, or omissions with regard to the proposed commission policy:

If the dollar amount involved in the order is less than $1,000, then the base commission will be 8.4 percent of the money involved. If the dollar amount involved in the order is at least $1,000 but less than $10,000, then the base commission will be 5 percent of the money involved plus $34. If the dollar amount involved in the order is at least $10,000, then the base commission will be 4 percent of the money involved plus $134. Regardless of the dollar amount involved in the order, the price per share and the number of shares of the stock or warrant also have an effect on the total commission. (Total commission is equal to the base commission plus surcharges figured on the price per share and the number of shares in the transaction.) On shares selling for under $14 per share, the total commission is equal to the base commission plus 5 percent of the base commission, unless the number of shares bought or sold is not a multiple of 100 shares, in which case the total commission is equal to the base commission plus 9 percent of the base commission. On shares selling from $14 per share to $25 per share, the total commission is equal to the base commission plus 2 percent of the base commission, unless the number of shares bought or sold is not a multiple of 100 shares, in which case the total commission is equal to the base commission plus 6 percent of the base commission. (Shares bought or sold in other than multiples of 100 shares, that is, odd lots, result in a 4 percent surcharge on the base commission.)

1. Identify all conditions (not resulting actions) in the problem. Create data element (DE) and data element value (DEV) definitions for all conditions in the problem, as shown below in Table 6.1.

Table 6.1
DEs and DEVs for Problem Conditions

Data Elements (DEs)	Data Element Values (DEVs)
1. DOLLAR–AMOUNT	S Small (0–$999.99)
	M Medium ($1,000–$9999.99)
	L Large (at least $10,000.00)
2. PER-SHARE-COST	C Cheap (zero–$13.99/share)
	A Average ($14–$25/share)
	X Expensive (above $25/share)
3. TYPE-LOT	E Even (multiples of 100 shares)
	O Odd (not multiple of 100 shares)

2. Establish the total number of combinations of conditions in the problem by multiplying the number of DEVs within each unique DE by each other. See Fig. 6.13.

There are 3 DOLLAR-AMOUNT data element values (S, M, and L).

There are 3 PER-SHARE-COST data element values (C, A, and X).

There are 2 TYPE-LOT data element values (E and O).

Therefore:

$$3 \times 3 \times 2 = 18$$

(Total number of combinations of conditions in the problem)

Figure 6.13 **Determination of the total number of problem conditions.**

3. Identify each of the independent actions in the problem. This is depicted in Table 6.2.

Table 6.2
Identification of Actions in the Problem

ACTIONS

8.4% base commission

5% + $ 34 base commission

4% + $134 base commission

5% of the base commission

2% of the base commission

4% of the base commission

4. Create the format of a decision table by numbering each of the combinations of conditions at the top of a diagram and by listing the DEs and each of the actions at the far left of the diagram, as demonstrated in Table 6.3.

Table 6.3
Decision Table Format Identified

	1	2	3	4	5	6	7	8	9	10	11	12	13	14	15	16	17	18
DOLLAR-AMOUNT PER-SHARE-COST TYPE-LOT																		
8.4% base 5% + $34 base 4% + $134 base 5% of base 2% of base 4% of base																		

5.   As shown in Table 6.4, fill in the decision table conditions (DEVs) for the top-most data element by applying the following formula:

$$\frac{\text{total number of combinations in problem}}{\text{number of conditions within current DE}} = \begin{array}{l}\text{repetition factor for}\\ \text{each condition (DEV)}\\ \text{within current DE}\end{array}$$

Table 6.4
Top Data Element Conditions Identified

There are 18 combinations in the problem. There are 3 conditions (S, M, and L) within the current data element.
Therefore:

$$\frac{18}{3} = 6$$

(where 6 is the repetition factor for each data element value within the current data element)

	1	2	3	4	5	6	7	8	9	10	11	12	13	14	15	16	17	18
DOLLAR-AMOUNT PER-SHARE-COST TYPE-LOT	S	S	S	S	S	S	M	M	M	M	M	M	L	L	L	L	L	L
8.4% base 5% + $ 34 base 4% + $134 base 5% of base 2% of base 4% of base																		

6. For each successive data element listing in the table, apply the following formula to fill in the conditions (DEVs) within a data element: Now look at Tables 6.5, below, and 6.6, on the following page.

$$\frac{\text{repetition factor for conditions within previously listed DE}}{\text{number of conditions within current DE}} = \begin{array}{l}\text{repetition factor for} \\ \text{each condition (DEV)} \\ \text{within current DE}\end{array}$$

Table 6.5
Second Data Element Conditions Identified

The repetition factor for conditions within the previously listed data element is 6. There are 3 conditions (C, A, and X) within the current data element.

Therefore:

$$\frac{6}{3} = 2$$

(where 2 is the repetition factor for each data element value within the current data element)

	1	2	3	4	5	6	7	8	9	10	11	12	13	14	15	16	17	18
DOLLAR-AMOUNT	S	S	S	S	S	S	M	M	M	M	M	M	L	L	L	L	L	L
PER-SHARE-COST	C	C	A	A	X	X	C	C	A	A	X	X	C	C	A	A	X	X
TYPE-LOT																		
8.4% base																		
5% + $34 base																		
4% + $134 base																		
5% of base																		
2% of base																		
4% of base																		

Table 6.6
Third Data Element Conditions Identified

The repetition factor for conditions within the previously listed data element is 2. There are 2 conditions (E and O) within the current data element.

Therefore:

$$\frac{2}{2} = 1$$

(where 1 is the repetition factor for each data element value within the current data element)

	1	2	3	4	5	6	7	8	9	10	11	12	13	14	15	16	17	18
DOLLAR-AMOUNT	S	S	S	S	S	S	M	M	M	M	M	M	L	L	L	L	L	L
PER-SHARE-COST	C	C	A	A	X	X	C	C	A	S	X	X	C	C	A	A	X	X
TYPE-LOT	E	O	E	O	E	O	E	O	E	O	E	O	E	O	E	O	E	O
8.4% base																		
5% + $34 base																		
4% + $134 base																		
5% of base																		
2% of base																		
4% of base																		

7. Identify the appropriate actions to be taken by analyzing each set of vertical conditions. See Table 6.7, on the following page.

Table 6.7
Identification of Decision Table Actions

	1	2	3	4	5	6	7	8	9	10	11	12	13	14	15	16	17	18
DOLLAR-AMOUNT	S	S	S	S	S	S	M	M	M	M	M	M	L	L	L	L	L	L
PER-SHARE-COST	C	C	A	A	X	X	C	C	A	A	X	X	C	C	A	A	X	X
TYPE-LOT	E	O	E	O	E	O	E	O	E	O	E	O	E	O	E	O	E	O
8.4% base	✓	✓	✓	✓	✓	✓												
5% + $34 base							✓	✓	✓	✓	✓	✓						
4% + $134 base													✓	✓	✓	✓	✓	✓
5% of base	✓	✓					✓	✓					✓	✓				
2% of base			✓	✓					✓	✓					✓	✓		
4% of base					✓	✓					✓	✓					✓	✓

8. Identify any omissions, ambiguities, or contradictions in the problem.

9. Create a list of questions to be answered by the users to clarify any omissions, ambiguities, or contradictions in the problem. See Fig. 6.14.

The narrative specifies that the price per share has an effect on the total commission. When the price per share is less than $14, a 5% surcharge is indicated. When the price per share is between $14 and $25, a 2% surcharge is indicated. No positive statement is made to identify the surcharge involved when the price per share is more than $25.

1. Should there be any surcharge when a share sells for more than $25?

2. If so, what is the surcharge?

Figure 6.14. Questions for user clarification.

10. Revise the decision table based on clarifications to the problem provided by the users.

11. Attempt to optimize the decision table by

(a) reducing the size of the table by eliminating or combining conditions according to common actions; and by

(b) specifying the order in which combinations of conditions should be tested, based upon relative frequency of conditions

6.4.2 Decision table variations

In the previous example, the decision table was filled in with DEVs defined by the problem narrative. This approach is particularly useful whenever there are more than two data element values within any data element in the problem. Other approaches to drawing decisions tables, however, frequently are used. Combinations of conditions can be depicted horizontally as well as vertically. Decision tables sometimes are filled with T/F (TRUE/FALSE) and Y/N (YES/NO) values instead of data element values. Occasionally, we may see decision tables with redundant sets of conditions. Such variations are shown in Tables 6.8 and 6.9, on the facing page, and reflect the policy stated in the following narrative:

If the product is selling well and stock is low, increase production of the product. If the product is selling well and stock is not low, continue production as is. If the product is not selling well and stock is low, continue production of the product as is. If the product is not selling well and stock is not low, discontinue production.

Table 6.8
Decision Table Filled with True and False Values

	1	2	3	4
PRODUCT-SALES-GOOD	T	T	F	F
INVENTORY-AMOUNT-LOW	T	F	T	F
INCREASE PRODUCTION	√			
CONTINUE PRODUCTION		√	√	
DISCONTINUE PRODUCTION				√

No matter which approach is used to draw a decision table, the analyst must verify that the problem has been stated clearly and properly. Just as well-stated objectives should be precise and measurable, well-stated conditions also must be precise and measurable. What is meant by the phrase, "if the product is selling well"? How do we know when we are low on stock? To answer these questions, the analyst must get clarification from the users.

Table 6.9
Decision Table Reflecting Redundancy and Y/N Values

CASE	PRODUCT-SALES GOOD	PRODUCT-SALES BAD	INVEN-TORY OK	INVEN-TORY LOW	ACTION
1	Y	N	Y	N	continue production
2	N	Y	Y	N	discontinue production
3	Y	N	N	Y	increase production
4	N	Y	N	Y	continue production

Decision tables are particularly useful when a disciplined approach is needed to organize the many conditions and actions in a problem. The more conditions and actions in a problem, however, the larger the decision table becomes. The problem examples in this chapter conveniently have been contrived so that decision table solutions would fit on one page. But what if the stock brokerage commission policy problem also included a commission differentiation based upon SOURCE-EXCHANGE (New York, American, Over the Counter, and other)? The number of combinations of conditions in the problem would increase from 18 to 72. As a result, the decision table would be too large to fit on one standard sheet of paper. It might not fit even on two sheets of paper. Since the usefulness of this kind of document is inversely proportional to its size, the analyst should consider reducing the size of the table.

Two approaches are recommended to do this: The first approach is quite simple — divide and conquer. Break one unmanageably large decision table into a few smaller decision tables. The second approach involves more thought by the analyst. After drawing a large decision table, the analyst should determine the individual conditions that combine to result in the same action. These conditions potentially can be redefined as a complex of conditions, thereby reducing the overall size of the table. The reader should note that this approach may present future problems if and when the actions for particular combinations of conditions change.

6.5 Decision tree structures

Tree structures have been advocated strongly for many years by data processing professionals as a tool to develop systems designs. Traditionally, these diagrams have been used to show the structure of data, systems, programs, and module logic. For example, Jackson[7] and Warnier[8] state that program structures should be based upon and directly related to the structure of input and output data. Jackson shows the data structure of a serial file in which detail records are arranged by district within division by means of a vertical tree structure, as adapted on the following page in Fig. 6.15. The asterisk notation in the upper right corner indicates that there are multiple records within a district, multiple districts within a division, and multiple divisions within a file.

Warnier achieves essentially the same result by using a horizontal tree structure with a bracket graphic notation (see Fig. 6.16, on the next page). Orr presents a further adaptation of the Warnier structure.[9] HIPO function charts and structure charts (discussed in more detail in Chapter 7) also are types of tree structures used to show the hierarchy of programs in a system as well as the hierarchy of modules in a program.

Tree structures can be used alternatively to show the conditions and actions in a logic problem. These decision tree structures are advantageous in that they are easy to create and involve no user training. The diagram is so simple that it essentially is self-explanatory.

In contrast, experiences with decision tables have been far less satisfactory. For example, even analysts who already are familiar with decision tables sometimes have difficulty drawing relatively simple decision tables. These problems are compounded whenever a data element has more than two data element values. In addition, users frequently require detailed explanations in order to understand decision table solutions. These problems can be eliminated by using decision tree structures instead of decision tables.

To see how a decision tree structure solution compares to a structured English solution and to a decision table solution, study Fig. 6.17, which presents a decision tree structure solution to the stockbrokers' commission policy problem.

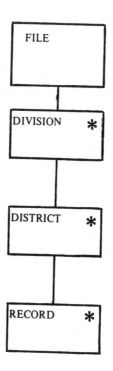

Figure 6.15. Vertical data tree structure using the Jackson approach.

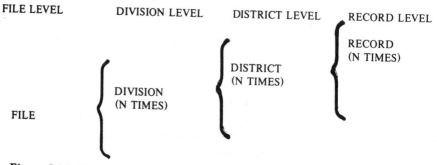

Figure 6.16. Horizontal data tree structure using the Warnier approach.

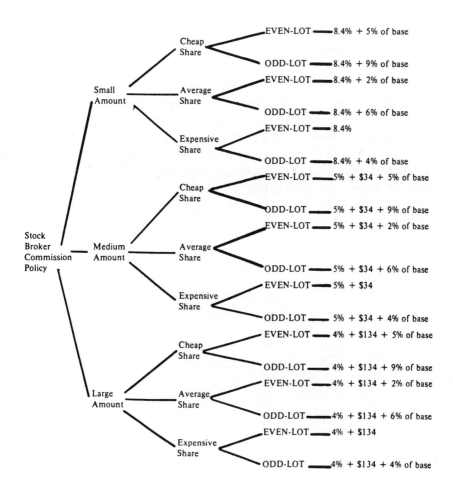

Figure 6.17. Decision tree structure showing broker commission policy.

The positive features of the decision tree structure should be obvious from the above figure. While we know that there are three types of dollar amounts (small, medium, and large), three types of per-share costs (cheap, average, and expensive), and two types of lots (even and odd), for many analysts, the transposition of this information into a decision table format is not trivial. Yet, almost all analysts can transpose this information into a decision tree structure without substantial difficulty. In addition, the decision tree structure is very readable, and is both self-explanatory and self-supporting.

Going back to the decision table problem solution shown in Table 6.7, analysts and users might have difficulty analyzing the decision table because of the abbreviated data element value notation. For instance, the answer to "What happens when a small amount of a cheap stock gets sold in an odd-lot?" is more apparent in the decision tree structure than in the decision table. The decision table would be more readable if it were supported by a legend of data element values and their abbreviations, but that would not eliminate all the problems associated with drawing a decision table.

Review Exercise

Consider the following policy:

The Swell Store employs a number of salesmen to sell a variety of items. Most of these salesmen earn their income from a commission, paid on the items they sell, but a few are salary-plus-commission employees — that is, they receive a fixed salary, regardless of the quantity or type of items they sell, plus a commission on certain items. The Swell Store sells several different lines of merchandise, some of which are known as standard items (a can of tomato soup, for example) because they are widespread and do not require any creative sales techniques; in addition, there are bonus items that are highly profitable but difficult to sell (a gold-plated, diamond-studded Cadillac, perhaps). The standard and bonus items generally represent the low and high ends of the price spectrum, sandwiching a greater number of items in the middle of the spectrum.

Customers, also, are categorized: Some are known as regulars, because they do business so often that no creative selling is required. Most of the customers, however, do a small amount of business at the Swell Store, and are likely to walk in right off the street, buy something, and then disappear forever.

The management's commission policy is as follows: If a non-salaried employee sells an item that is neither standard nor bonus to someone other than a regular customer, he receives a 10 percent commission, unless the item costs more than $1,000, in which case the commission is 5 percent. For all salesmen, if a standard item is sold, or if any item is sold to a regular customer, no commission is given. If a salaried salesman sells a bonus item, he receives a 5 percent commission, unless the item sells for more than $1,000, in which case he receives a flat $25. If a non-salaried salesman sells a bonus item to someone other than a regular customer, he receives a 10 percent commission, unless the item sells for more than $1,000, in which case he receives a flat commission of $75.

Utilizing this information, practice your skills as follows:

1. Define the data elements and the data element values associated with the commission policy.

2. Create structured English, a decision table, and a decision tree structure to depict the combinations of conditions and the resulting actions of the problem.

3. Identify all omissions, ambiguities, and contradictions in the policy.

References

1. "Software Management — Decision Table Processors," *ICP Quarterly*, October 1974, p. 111.

2. R.D. Carlsen and J.A. Lewis, *The Systems Analysis Workbook* (Englewood Cliffs, N.J.: Prentice-Hall, 1973).

3. "Computer Surveys," *ACM Journal*, December 1974.

4. H. McDaniel, ed., *Application of Decision Tables* (Princeton, N.J.: Brandon/Systems Press, 1970).

5. T.R. Gildersleeve, *Decision Tables and Their Practical Applications in Data Processing* (Englewood Cliffs, N.J.: Prentice-Hall, 1970).

6. S. Pollack, H. Hicks, and W. Harrison, *Decision Tables — Theory and Practice* (New York: John Wiley & Sons, 1971).

7. M.A. Jackson, *Principles of Program Design* (New York: Academic Press, 1975).

8. J.D. Warnier, *The Logical Construction of Programs*, 3rd ed., trans. B.M. Flanagan (New York: Van Nostrand Reinhold, 1976).

9. K.T. Orr, *Structured Systems Development* (New York: YOURDON Press, 1977).

7 Structure Charts

The structure chart provides the design reviewer with a document that serves as the focus of the design evaluation process.

7.1 Perspective

It often is suggested that a systems analyst should be responsible for producing the physical designs associated with a systems development effort. Strategies such as transform analysis and transaction analysis make the transition from a logical design to the first draft of a physical design a relatively straightforward process (see Chapter 9). Certainly, having one person do both the analysis of requirements and the physical design for a system minimizes potential communication problems that result in distortion and loss of detail. Yet, frequently, the complexity of the systems development effort goes beyond the technical expertise of the systems analyst.

7.2 Traditional approaches

Organizations vary in their approaches to the problem. In many small and moderate size organizations, programmer/analysts are responsible for the entire systems development effort. Sometimes, analyst/designers are responsible for all systems development work, from feasibility study through physical design. Other organizations have separated the functions of analysis and physical design so that an analyst does only the analysis phase, and the physical designer does only physical design. This third approach is especially useful when complex systems development efforts require considerable specialization of function. But in any approach, these systems developers call upon experts on an as-needed basis to fill the technical gaps.

Regardless of the approach taken to create a physical system's design, the systems analyst is responsible for assuring that the users' objectives have been met. Therefore, prior to the physical design phase, the analyst and the users should agree on the functional requirements and constraints of the proposed system: Both parties should know what the system is supposed to do, the speed at which it is to do it, the estimated volume to be processed, and the approximate cost of the project.

Clearly, how the logical design is physically implemented will affect whether the users' objectives are met. Since users generally do not have the time or are not technically competent to review physical system's designs, this burden falls, by default, on the systems analyst. The analyst actively must evaluate the physical design of a system to answer the following questions:

- Does the physical design match the logical functional requirements?

- Can the system perform within the specified time constraints?

- Can the system handle the anticipated volume of processing?

- Can the physical system be produced within the established time frame and cost limitations?

- Is the physical system's design comprised of independent, functional modules so that it will lead to relatively trouble-free and inexpensive maintenance?

- Will the system meet the users' expectations?

Unfortunately, these questions cannot be answered by studying the typical documentation of a physical system's design. For example, let's take a look at common forms of physical design documentation to determine how completely they answer our questions. Figure 7.1 depicts a physical version of a data flow diagram in the form of a system's I/O flow, but it will not answer our questions. The diagram *is* useful in identifying the physical devices that are the sources and destinations of the data in the system. We also can envision that transactions are edited, files are accessed and updated, statistics are taken, and reports are produced. But we know nothing about the structure and function of the modules in the system. We cannot evaluate whether the system will meet the functional objectives of the users, how big it is, how long it will take, or how easy it will be to maintain.

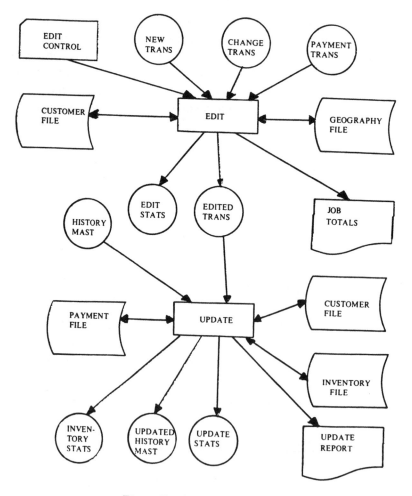

Figure 7.1. Systems I/O flow.

Table 7.1
System's File Description

FILE ID	FORMAT	LENGTH	BLOCK SIZE	FILE ORG.	VOLUME	GROWTH
FILE1	U	1000	1000	PS	10,000	15%
FILE2	F	600	6000	IS	1,000	8%
FILE3	F	600	6000	DA	1,000	10%
FILE4	V	124	3000	PS	Negligible	15%
FILE5	F	400	4000	PS	20,000	30%

Take a look at other examples. Table 7.1, above, describes files in tabular format and addresses anticipated growth in data volume, but it does not address system's functions, costs, or maintenance concerns. The module overlay structure in Fig. 7.2 gives the systems analyst some idea of the physical organization of the modules, but their functions, relationships, and independence are not clear. Figure 7.3 is an executive flowchart that establishes the sequence of general functions; it does provide an overview of the system's processing, but supplies no information to confirm the detailed functions of the system.

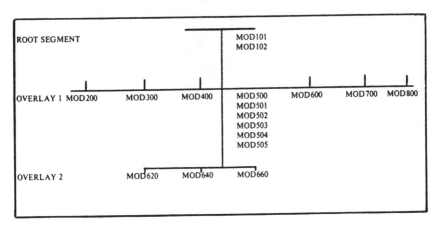

Figure 7.2. Module overlay structure.

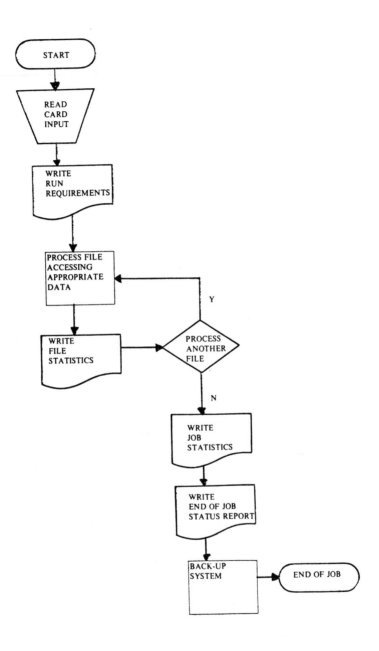

Figure 7.3. Executive system's flowchart.

These typical physical design documents normally would be supplemented by more detailed flowcharts, physical data record layouts, and narratives about the system, its programs, and its modules. In total, they comprise a traditional systems design documentation package. The point is, however, that this package does not indicate clearly that the physical design meets the objectives of the users. Neither the rambling narrative specifications nor these graphics convey that functional specifications and time and cost limitations can be met.

7.3 Function charts

Many organizations have recognized that there are serious weaknesses in traditional physical system's design documentation. IBM, in its HIPO documentation technique (*H*ierarchy plus *I*nput-*P*rocess-*O*utput), uses a graphic approach to identify the inputs, outputs, functions, and modular hierarchy of a system.[1,2,3] The HIPO table of contents, also known as a function chart and depicted in Fig. 7.4, provides the analyst with a graphic representation of the interrelationships of the modules in the system. As first developed, however, the function chart did not indicate clearly the function of each of the modules. Look closely at Fig. 7.4. Does the GROSS PAY module compute gross pay only, or does it do gross pay editing as well? Does the PAYCHECKS module only generate paychecks, or does it also determine whether a paycheck should be generated? Recognizing that function chart modules contained functional ambiguity, IBM suggested the use of a functional verb as part of the module name. The enhanced function chart is shown in Fig. 7.5.

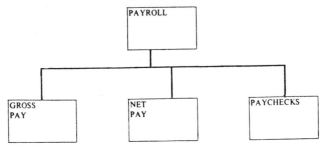

Figure 7.4. Function chart.

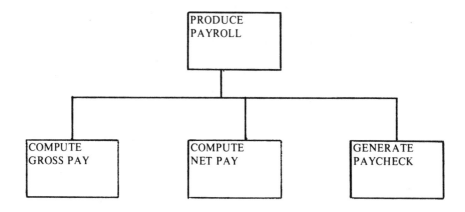

Figure 7.5. Enhanced function chart.

A function chart, especially one in which modules are described by a verb and object, is a more useful tool for the analyst than other previously described physical design documents. It shows the modular hierarchy of the system as well as modular partitioning and functions — but it still does not show data and control communication in the modular hierarchy, nor major loops or decisions involved in processing. This additional information may be included in the detailed IPO diagram (see Fig. 7.6), but its relationship to the overall physical hierarchy of modules still may be unclear. For example, by reading a detailed narrative for the processing of the EDIT-TRANSACTION module of a system, we might determine that a major decision is made in that module. But in how many other modules is this same decision made? Normally, we would have to read the narratives or study the detailed IPO diagram of every module in the system to determine the answer. Or, in reading the narrative of the EDIT-TRANSACTION module, we might find that the module needs a transaction in order to do its job. How can the analyst be assured that the transaction is available at the time the module is invoked? Where did the transaction originate? Which module passed it to which other modules? These questions cannot be answered by looking at a function chart.

From Module MAIN

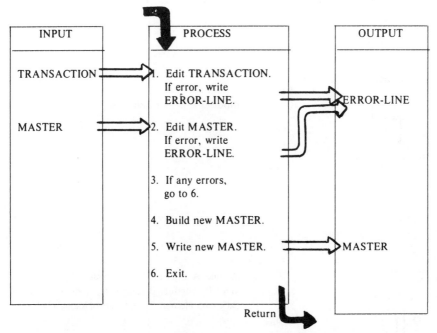

Figure 7.6. Detailed IPO diagram for module EDITANDUPDATE.

7.4 Structure charts

A systems design graphic that remedies the weaknesses associated with function charts is the structure chart. In use since the late 1960's and formally introduced to data processing professionals in the classic paper entitled "Structured Design," by Stevens, Myers, and Constantine,[4] this graphic superimposes onto a function chart format the data and control communication required for modules to work. By distinguishing data from control, a structure chart clearly indicates intermediate work areas and switches present in a system. A structure chart also may show major loops and major decisions in a system. *Above all, it provides the design reviewer with a document that serves as the focus of the design evaluation process, so long as all program and module names, functions, and interfaces are defined rigorously, preferably in a data dictionary* (see Chapter 8).

a module named EDIT-XACT

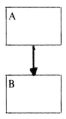

module A calls module B

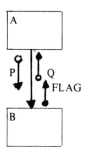

module A calls module B passing to it data element P. Module B returns to module A sending back data element Q and control element FLAG.

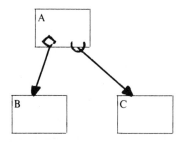

module A calls module B on the basis of a major decision in module A. Module A calls module C on the basis of a major loop in module A.

Figure 7.7. Structure chart symbols.

Figure 7.7 shows the symbols used in a structure chart.* Each module is represented by a rectangle identified by a functional module name. The module name (or alias, if constrained by organizational module-naming conventions) should be comprised of a descriptive verb and, optimally, a single, nonplural object identifying the total function of the module. Two modules are connected by an arrow that is drawn from the edge of one module rectangle to the edge of the other. Although the arrow head points in only one direction from the calling to the called module, it is understood that the called module returns to the calling module. When one module calls another, the calling module may send data or control information to the called module so that it can operate. The called module may produce data and/or control information, which is passed back to the calling module. This required data and control communication between modules is shown along the arrow indicating the modular connection. Note that data communication is shown by an arrow with an open circle, and control communication by an arrow with a solid circle.

The call from one module to another may result from a major decision or a major loop in the system. Such conditional calls are noted respectively by the diamond and semicircle, as shown in Fig. 7.7. Notations of decisions and loops on the structure chart should represent only the most critical decisions and loops in the system to minimize cluttering the diagram.

The PAYROLL subsystem or program, pictured by a function chart in Fig. 7.5, is shown as a structure chart in Fig. 7.8. The structure chart not only illustrates the hierarchical modular structure, but also indicates clearly each module's function and the communication required between modules. The systems analyst can see from this one diagram whether the physical

---

*Structure chart symbology sometimes includes reference to so-called pathological connections when one module references data defined in another module, or when the called module does not return to the calling module. This type of modular connection normally is so undesirable that no structure chart symbology is presented to support its existence. For detailed treatment of pathological connections, see Edward Yourdon and Larry L. Constantine, *Structured Design: Fundamentals of a Discipline of Computer Program and Systems Design,* 2nd ed. (New York: YOURDON Press, 1978), pp. 235-249.

design matches the logical functional requirements. Although the structure chart does not indicate how fast the system will run or how much volume it can process, it does provide the analyst with considerable information. For example, the analyst can evaluate the functional independence of the modular design, how long the programming implementation should take, and how easy the system will be to maintain.

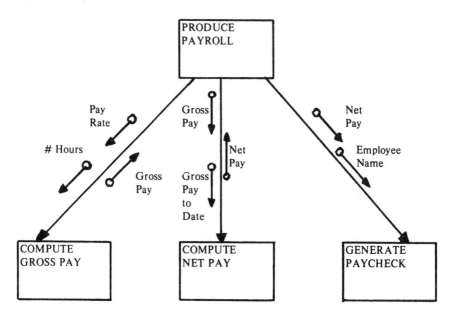

Figure 7.8. Structure chart for the PAYROLL program.

   If one were to track the sequence of events depicted in Fig. 7.8, it would appear that gross pay is computed before net pay, and that the paycheck is generated next. One might infer from the figure that a structure chart indicates the sequence in which modules are invoked. High-level modules, of course, must be invoked before they can invoke connecting, lower-level modules. However, the horizontal placement of modules in a structure chart does not indicate any particular sequence of execution; Fig. 7.9 shows two structure charts that are logically and physically equivalent, although they do not have the same horizontal order of modules. Note also that module E — although referenced by two different modules, B and C — appears only once in each rendering of the structure chart.

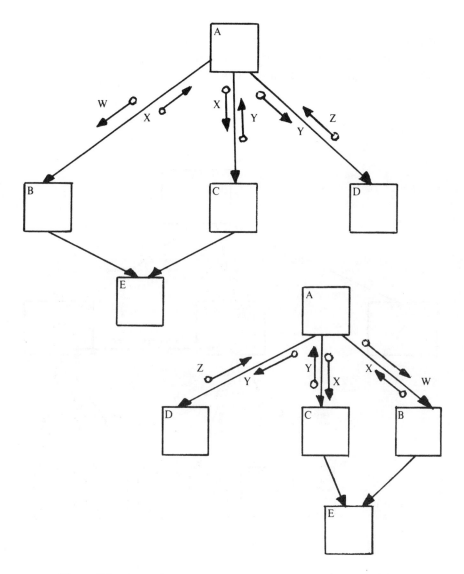

Figure 7.9. Logically and physically equivalent structure charts.

By seeing major decisions and major loops symbolized on a structure chart, the analyst more easily can evaluate the organizational efficiency of the modular structure. Figure 7.10, for example, indicates that computing net pay requires that a decision on employee type be made twice. Duplicate decision-making frequently is an indication of poor design. While strategies to correct such a problem will not be suggested until Chapter 9, it is important to note here that when major decisions and major loops are noted carefully on a structure chart, the analyst easily can perceive potential design flaws.

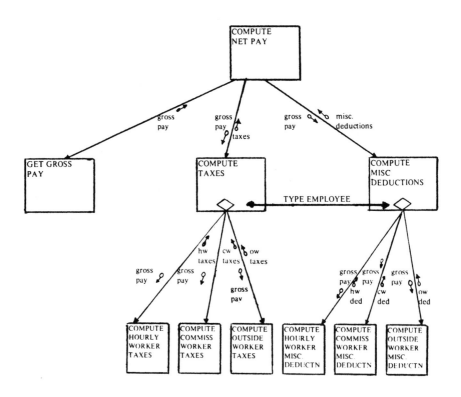

Figure 7.10. Structure chart with duplicate decision-making.

This chapter has identified some of the ways in which structure charts enable improved systems design and analysis. Before leaving this subject, let us consider one more way in which the structure chart functions: It facilitates change. Because it clearly identifies inputs, outputs, and processes, one should not have to delve into the program listings to identify the modules affected by a required change. Of course, module listings will have to be examined inevitably, but only to determine *how* to make the change, not *whether* or *where* a change is necessary. In this way, just as the data flow diagram acts as a blueprint for the system's logical design, the structure chart is the blueprint of the system's physical design.

Review Exercise

1. What documentation would you use to evaluate the physical design of a system if that documentation did not include a structure chart?

2. What criteria would you use to evaluate the physical design of a system?

3. What is the difference between the following modular structures?

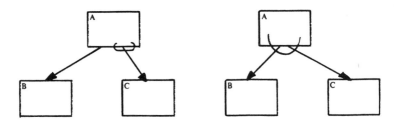

4. Interpret the following diagram. How many times is module C called? Is module B called before module C?

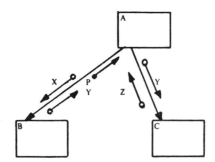

5. How would a data dictionary be helpful to support a structure chart as a rigorous design communication tool?

References

1. *HIPO: A Design Aid and Documentation Technique,* Form GC20-1851-0, IBM Corporation, 1974.

2. M.N. Jones, "HIPO for Developing Specifications," *Datamation,* Vol. 22, No. 3 (1976), pp. 112-114, 121, 125.

3. H. Katzan, Jr., *System Design and Documentation: An Introduction to the HIPO Method,* (New York: Van Nostrand Reinhold, 1976).

4. W. Stevens, G. Myers, and L. Constantine, "Structured Design," *IBM Systems Journal,* Vol. 13, No. 2 (1974), pp. 115-39.

5. E. Yourdon and L.L. Constantine, *Structured Design: Fundamentals of a Discipline of Computer Program and Systems Design,* 2nd ed. (New York: YOURDON Press, 1978), pp. 235-249.

8 Data Dictionary

Unfortunately, systems developers are in such a hurry to get to the solution that they frequently skip some of the intermediate steps that will help assure successful systems.

8.1 Perspective

Systems development involves the accumulation of a vast amount of information which, if not precisely defined, can subvert systems development efforts. Problems arise when the same piece of information or even the same word has different meanings, depending on how and in what context it is used. As an example, we might say that a telephone number is comprised of an area code, an exchange, and a line number with, perhaps, an extension. Were we to use these terms at a business meeting within a telephone company, those present would understand our terminology. However, would we understand what they meant if they used the terms national plan area and central office, instead of area code and exchange, respectively? Or, at a technical meeting within the EDP department of a telephone company, would we understand such terms as ONPA (originating area code) and OCO (originating exchange)? The point is not so much that different people speak different languages, but rather that some people understand only their own language. If successful systems are to be built, business and technical people must communicate clearly with each other.

The problem of precise definition goes beyond terminology. Automated systems development involves the reduction of procedures and policies into program code, as well as the intermediate steps of logical and physical systems design. The policies and procedures will be unclear unless terms are defined; the logical design of the system, expressed in data flow diagrams and data structure diagrams, will be incomprehensible unless data flows, processes, entities, and attributes are defined; and the physical design of the system will not be reviewable unless programs, modules, subroutines, and the associated data and control linkages are defined with precision.

Unfortunately, systems developers are in such a hurry to get to the solution that they frequently skip some of the intermediate steps that will help assure successful systems. Often, management contributes to the problem by forcing systems analysts to develop premature cost estimates and schedules that are based on an incomplete understanding of the problem and potential solutions. While developers rush to produce some real output, systems documentation, a critical systems development activity, typically is deferred until a system has been almost totally tested and is ready to be accepted and installed. At that point, documentation becomes a pure chore, a necessary evil (defined by standards), which further delays the system's acceptance when the system already may be late. As a result, the documentation effort often is rushed, and the skimpiest documentation package that will meet the standards of the organization is delivered. The usefulness of such systems documentation, even after user acceptance, may be suspect.

8.2 Data dictionary objectives

Just as a dictionary defines the words in a language, a data dictionary defines terms associated with systems development. A data dictionary (also known as a data directory, data catalog, element catalog, or data encyclopedia) by itself will not assure improved systems documentation. Normally, however, the creation and use of a data dictionary is supported by forms and procedures (discussed in Sections 8.3 and 8.4) to deal with the flood of detail generated by the developmental process. These procedures, whether manual, automated, or some combination of the two, help to assure timely, complete, and precise definitions of terms as they are being used, not after the fact. As a result, a system is documented as it is developed. Since that documentation includes definitions of all terms in data flow diagrams, data structure diagrams, and structure charts, our design graphics become rigorous tools for specification.

To understand the objectives of a data dictionary, consider the following situation:

> An executive in New York and an executive in Los Angeles both want to access information from the organization-wide data base.

A data dictionary, or at least some form of documentation defining all of the data elements in the data base, provides a standard terminology so that both executives know how to define their requirements.

Here's another situation:

> The super-duper programmer has quit and his super-duper program has just crashed. The new maintenance programmer, probably a trainee, is looking at a module listing, which has no comments, and cannot figure out what the module does.

A data dictionary frequently is thought of as a repository of data about data.[1] As such, it could be used as a reference that defines data elements manipulated in program or module logic. A data dictionary, however, also could include definitions of both logical and physical processes. While there is no defense for an unmaintainable module listing such as the one in the above example, a data dictionary could describe the logic both from the users' point of view (logical processes) and from a programmer's perspective (physical processes).

Consider the following:

> A programmer is told to write a routine to validate a date that is input to the system in MM/DD/YY format. After a week of coding and testing, the programmer delivers the routine only to find out that a routine already exists to validate a date in MMDDYY format.

A data dictionary, especially one with good naming conventions, can be used as a reference to identify predefined modules available to all programs in the organization. Reference to such a dictionary in the above case would have resulted in no duplication in programming and, perhaps, standardization of the date interface (i.e., elimination of the slashes in the MM/DD/YY format).

Some other examples:

- The Social Security Administration has discovered that within a few years there will be more people covered by Social Security than can be identified by a nine-digit number.

- The State Motor Vehicle Department has discovered that within a few years there will be more drivers than can be identified by a seven-digit driver's license number.

- The PDQ Company has decided to change its account identification to reflect the geographic location of all customers.

- The Alaska pipeline is now sending more barrels of oil per day from one source than the programs in the system can handle.

In all of these situations, programs in the related systems will have to change to solve the problems. In a case similar to the third case above, a major corporation recently spent many man-years in a systems redesign effort because their current systems documentation did not identify where to make changes.

In a related case of new oil from Alaska, another major corporation investigated every program in the system to identify where the number-of-barrels-shipped-per-day field was manipulated. After considerable time and effort, the analysts assigned to the task identified more than eighty places where the programs had to be changed. (Do you think the analysts would bet on there not being another change?) In all of the cases, a data dictionary with a cross-reference capability could be used to identify all system's modules in which a data element is manipulated.

Let's look at another example:

- The programmers have finished coding modules and are ready to test them. Since the users haven't had time to create any test data, the programmers are told to create the appropriate test data.

This is not a hypothetical case: The generation of test data is required by any automated system. The activity normally is tedious, time-consuming, and prone to error in the sense that incomplete test data results in partially untested systems. A data dictionary that defines all data elements and their interrelationships could be used as an aid in generating test data and in verifying proper test results. Figure 8.1 represents this.

DATA ELEMENT TYPE	ALLOWABLE VALUES	TEST CASES GENERATED	STATUS	
			valid	invalid
STATUS-CODE (Numeric Range)	100–699	399 (middle value)	√	
		699 (high value)	√	
		100 (low value)	√	
		700 (too high)		x
		099 (too low)		x
		000 (zero)		x
		(blank)		x
		ABC (alphabetic)		x
		5×2 (alphanumeric)		x
DEPT (Discrete Values)	SALES ACCTG MRKTG PUBLG	SALES (all defined)	√	
		ACCTG	√	
		MRKTG	√	
		PUBLG	√	
		ABCDE (undefined alpha)		x
		(blank)		x
		12345 (numeric)		x
		AB5DE (alphanumeric)		x

Figure 8.1. Data dictionary for test data generation.

Suppose you were faced with the following situation:

Today's date is November 15, 1999. All date validation logic in every program in every system of your
organization must be investigated to assure that the
year 2000 is *not* considered a leap year.

A data dictionary could be used to identify all modules that handle date validation logic. If all programs utilize a standard date
validation routine (defined in the data dictionary), then a change
only to that standard module must be made. Any modules with
in-line, nonstandard date validation logic also would be identified, thus minimizing potential maintenance problems.

Let's consider a situation in which a data element in a system must be changed:

CUST-STATUS, with four allowable values, will have
five allowable values as a result of the change. Charlie, the COBOL programming-change-coordinator, tells
Joe, the COBOL systems-maintenance-programmer, to
make the appropriate changes. Joe does a perfect job,
but two days later the system starts producing invalid
output as a result of the changes. Joe checks each of
his changes but cannot find the problem. Finally, he
realizes that a few modules in the system are not
written in COBOL. The problem, he realizes, is that
these other modules were not updated to reflect the
change in the CUST-STATUS data element.

A data dictionary could have helped to eliminate the above problem in a variety of ways. First, a data dictionary should completely identify synonyms, acronyms, and aliases of each data
element in the system, thereby warning Joe, our maintenance
programmer, that other changes may be required. In the above
case, for example, CUST-STATUS also is known (and used) by the
label CUSTSTAT. Second, reference to the CUST-STATUS data element should identify all of the modules in the system that use

that data element. As a result, the maintenance programmer would have known that some modules outside of his responsibility also might be affected by the change. Third, the data dictionary is the repository of the detailed documentation of the system. If all requests for changes are funneled through one set of rigorous procedures associated with the data dictionary, centralized control for changes will result, dramatically reducing maintenance costs.

In summary, a data dictionary supported by a rigorous set of procedures will help us achieve the following objectives:

- to establish a glossary of terms
- to provide standard terminology
- to define all terms associated with a system
- to identify modules available to all systems
- to provide cross-reference capability
- to help in the generation of manageable, complete test data
- to resolve problems associated with aliases and acronyms
- to provide a centralized control for systems changes
- to provide a reference guide for training and design evaluation
- to help minimize maintenance costs

8.3 Data dictionary definitions

The mechanism for defining data dictionary entries depends upon the type of support available. As an example, automated data dictionary packages include preprinted forms that indicate what can be defined and how to define it. Data dictionary systems developed by a particular organization for its own use will prescribe procedures for data definition to meet the specific needs of the organization. Irregardless of the approach used to document the detailed characteristics of systems and programs, all definitions should be accessible by name and organized either alphabetically or alphabetically within type of definition. Because data dictionary definitions are subject to change, they should be easy to revise, a task that can be accomplished readily if redundant definition is minimized.

Data dictionary definitions should reflect the documentation tools produced by the systems development process. A direct correlation should exist between the definitions in the data dictionary and the data flows and processes in data flow diagrams; the entities, records, and attributes in data structure diagrams; and the data communication and processes in structure charts. And, if the data dictionary is to be an effective tool for maintenance, all definitions should be cross-referenced to associated files and processes.

The specific characteristics of data dictionary definitions vary by type of definition. All definitions should include a concise description of the term, identification of associated aliases, and identification of the composition of the entry being defined. Data element definitions should identify allowable values, editing criteria, associated modules, and associated files. Data flow definitions should identify sources and destinations of data flows as well as pertinent characteristics such as volume, peaks, and valleys. Process definitions should identify all required inputs, outputs, and data base accesses, references to graphic communication tools (structure chart and data flow diagram reference numbers), and a process description that is expressed or supported by decision tables, decision tree structures, or structured English. The format of these definitions varies according to the type of information, the support facilities available, and the standards of the organization.

Figures 8.2 through 8.6 show these definitions, each with a separate form. Figure 8.7 is a suggested form for manual data dictionary systems, which allows the definition of all types of entries to be made on one form. Figure 8.8 provides an extract of a glossary of data terms, showing data composition using logical operands.

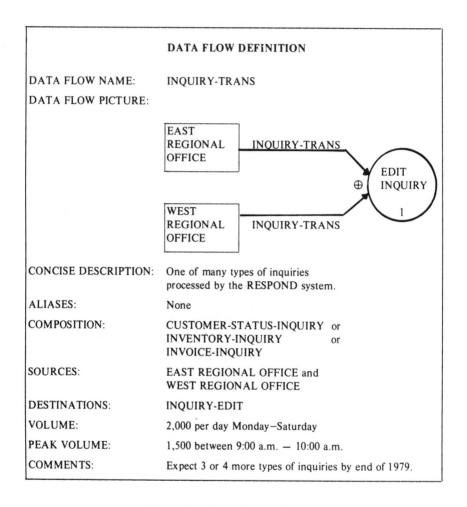

Figure 8.2. Data flow definition.

```
+------------------------------------------------------------------+
|                  DATA ELEMENT DEFINITION                         |
|                                                                  |
| DATA ELEMENT NAME:   DATE-ACCOUNT-OPENED                         |
| CONCISE DESCRIPTION:   Start date of a customer's account.       |
|                                                                  |
| ALIASES:                 OPENDATE                                |
| COMPOSITION:              MONTH-ACCOUNT-OPENED    and            |
|                           DAY-ACCOUNT-OPENED      and            |
|                           YEAR-ACCOUNT-OPENED                    |
|                                                                  |
| ASSOCIATED RECORDS, FILES, OR DATA BASES:                        |
|                           CUSTOMER-MASTER                        |
|                           CUSTOMER-STATUS-RESPONSE               |
|                           CUSTOMER-TRANS                         |
| ASSOCIATED PROCESSES:                                            |
|     PROCESS NAME              DFD REFERENCE     SC REFERENCE     |
|     EDIT-CUSTOMER-TRANS            3                14           |
|     BUILD-CUSTOMER-MASTER          6.2              18           |
|     BUILD-CUSTOMER-STATUS-RESPONSE 8.4              21           |
|                                                                  |
| DATA CHARACTERISTICS: NO. CHARACTERS  6   TYPE  N               |
| ALLOWABLE VALUES /RANGES:   MMDDYY format.  See                  |
|          MONTH-DAY-TABLE.  Year must not precede 42.             |
|          Date must be earlier than RUN-DATE.                     |
+------------------------------------------------------------------+
```

Figure 8.3. Data element definition.

PROCESS DEFINITION

PROCESS NAME: EDIT-CUSTOMER-TRANS

PROCESS ID NO: DFD 3

PROCESS PICTURE:

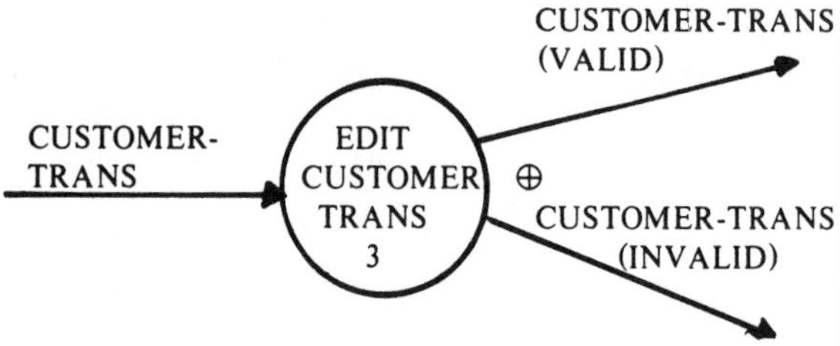

CONCISE DESCRIPTION: Process determines the validity of
the CUSTOMER-TRANS. No change is
made to the raw data as a result of editing.

ALIASES: None

INPUT DATA FLOWS: CUSTOMER-TRANS

OUTPUT DATA FLOWS: CUSTOMER-TRANS (VALID and INVALID)

DATA BASE ASSESSES:

DB NAME	ACCESS INFO	DATA RETRIEVED
_____	_____	_____
_____	_____	_____
_____	_____	_____

ASSOCIATED PROCESSES: SC 14

PROCESS LOGIC: See editing criteria for data elements comprising
CUSTOMER-TRANS. See structured English
attachment for interfield editing criteria.

Figure 8.4. Process definition.

```
┌─────────────────────────────────────────────────────────────────────┐
│                   FILE OR DATA BASE DEFINITION                        │
│                                                                       │
│ FILE □ or DATA BASE □ NAME:  EMPLOYEE                                  │
│                                                                       │
│ CONCISE DESCRIPTION:    Contains all information on                   │
│                         full time employees.                          │
│                                                                       │
│ ALIASES:                None.                                         │
│                                                                       │
│ COMPOSITION:            EMP-NAME              and                     │
│                         EMP-ID-NO             and                     │
│                         EMP-START-DATE        and                     │
│                         EMP-SALARY            and                     │
│                         EMP-REVIEW-DATE       and                     │
│                         EMP-DEPARTMENT        and                     │
│                        ⎧ PROJECT-ID-NO and  ⎫                         │
│                        ⎩ PROJECT-MANAGER ⎭    from 0-3                │
│                                                                       │
│ ORGANIZATION:           Sequential by EMP-ID-NO.                      │
│                                                                       │
│ ASSOCIATED PROCESSES:                                                 │
│                                                                       │
│    PROCESS NAME       DFD REFERENCE        SC REFERENCE               │
│                                                                       │
│    UPDATE-EMPLOYEE        7.1.2                46                     │
│    RETRIEVE-EMPLOYEE      7.2.4                51                     │
│                                                                       │
└─────────────────────────────────────────────────────────────────────┘
```

Figure 8.5. File or data base definition.

```
┌─────────────────────────────────────────────────────────────────────┐
│                        ALIAS DEFINITION                               │
│                                                                       │
│ ALIAS NAME:           OPENDATE                                        │
│                                                                       │
│ ALIAS TYPE:           Data element                                    │
│                                                                       │
│ SYNONYMS:             DATE-ACCOUNT-OPENED                             │
│                                                                       │
│ CONCISE DESCRIPTION:  Start date of a customer's account.            │
│                                                                       │
│ ALIAS ENVIRONMENT:                                                    │
│                                                      SYSTEM           │
│        CONTACT        DEPARTMENT                                      │
│                                                                       │
│        P. Smith       Systems Planning            FORECASTING        │
│                                                                       │
│ COMMENTS·             OPENDATE exists in a BAL program in the         │
│                       FORECASTING system.  That system is currently  │
│                       being rewritten in COBOL.  The approximate date│
│                       to discontinue supporting OPENDATE is June 1979.│
│                                                                       │
└─────────────────────────────────────────────────────────────────────┘
```

Figure 8.6. Alias definition.

```
┌─────────────────────────────────────────────────────────────────────┐
│ DATA      │                                                           │
│ DICTIONARY│   NAME: _____  │
│ ENTRY     │                                                           │
│───────────┘   TYPE:  DATA FLOW   DATA ELEMENT   PROCESS               │
│               (circle one)   FILE    DATA BASE    ALIAS               │
│                                                                       │
│ CONCISE DESCRIPTION: _____   │
│                                                                       │
│ ALIASES AND ABBREVIATIONS: _____   │
│───────────────────────────────────────────────────────────────────── │
│ IF ENTRY IS DATA ELEMENT:                                             │
│                                                                       │
│          NO. CHARACTERS _____ TYPE _____ REFERENCES ____        │
│          COMPOSITION        ALLOWABLE VALUES/RANGES                   │
│          _____    _____                  │
│          _____    _____                  │
│                                                                       │
│───────────────────────────────────────────────────────────────────── │
│ IF ENTRY IS DATA FLOW:                                                │
│          SOURCES:     _____      DESTINATIONS:_____               │
│                                                     _____          │
│          VOLUME:      _____      PEAKS:          _____          │
│          COMPOSITION:    _____   │
│                          _____   │
│                          _____   │
│───────────────────────────────────────────────────────────────────── │
│ IF ENTRY IS PROCESS:                                                  │
│          INPUTS: _____        OUTPUTS:_____             │
│                  _____        _____            │
│          DATA BASE ACCESSES:_____   │
│          ASSOCIATED PROCESSES:_____   │
│          PROCESS LOGIC ATTACHED:        YES ___NO ___                 │
│───────────────────────────────────────────────────────────────────── │
│ IF ENTRY IS FILE OR DATA BASE:                                        │
│          ORGANIZATION:_____ ASSOCIATED PROCESSES:_____        │
│          COMPOSITION:_____   │
└─────────────────────────────────────────────────────────────────────┘
```

Figure 8.7. Data dictionary entry.

```
INVOICE-INQUIRY          =   * Request for information about one customer invoice*

                         =   CUSTOMER-NAME  +
                             CUSTOMER-ADDR  +
                             INVOICE-NO

INVOICE-RESPONSE         =   * Identification and characteristics of
                             INVOICE in response to an INVOICE-INQUIRY*

                         =   CUSTOMER-NAME  +
                             CUSTOMER-ADDR  +
                             INVOICE-NO    +
                            ⎧ PRODUCT-NAME  + ⎫
                            ⎪ PRODUCT-ID-NO  + ⎪
                            ⎨ PRODUCT-PRICE  + ⎬   +
                            ⎩ SUB-ORDER-AMT    ⎭

                             TOTAL-INVOICE-AMT  +
                             AMT-PAID          +
                            ⎡ AMT-DUE                ⎤
                            ⎣ REFUND-AMT + STATUS ⎦   +
                             (INVOICE-COMMENTS)
```

Legend: =   means   is composed of
           +   means   and
          [ ] means   choose one of (exclusive or)
          ⟨ ⟩ means   at least one of (inclusive or)
          ( ) means   optional
          { } means   iterations of
          * * means   comment

Figure 8.8. Data definition using logical operands.

There is no shortcut to good systems documentation. It requires complete and precise definition of systems details, supported by considerable clerical work. Early in a systems development effort, the task of documentation often seems an ever-expanding nightmare, for the amount of detail can be overwhelming. Yet, early and ongoing control of this detail is essential to cost-effective systems development. As stated before, almost all systems terms should be defined prior to implementation so that users, designers, and programmers understand the specifications; subsequent systems enhancements and maintenance will necessitate continuous updating of documentation.

8.4 Data dictionary support resources

Two independent, but mutually supportive, resources can be utilized to help control the documentation process in general, and the data dictionary in particular. The first, in response to the need for clerical help, is either the data base administrator or the librarian, or sometimes both, depending on the size and number of projects supported by the data base and data dictionary. Either or both of these people have responsibility for the data dictionary's consistency and integrity. Typical duties include coordinating changes, updating the data base or data dictionary, and maintaining current listings. These roles are advisable no matter if supporting procedures are manual or automated.

The second resource is an automated data dictionary package.* There are many such aids to systems development and its associated documentation activities, and most have been designed to meet a variety of objectives.[2] In general, data dictionary packages support organization-wide data bases and general inquiry facilities; they provide control during systems development and maintenance phases; they aid in coordinating projects that extend beyond departmental and regional boundaries; and they provide an excellent reference for training.

Automated data dictionary packages thrive on redundancy. Terms that are merely unnecessary aliases of other terms can be detected and eliminated easily, providing more maintainable systems. The redundancy implicit in definitions which cross-reference other definitions help to assure integrity and consistency in the data dictionary. The overlap in process definitions facilitates automatically generated systems graphics. In effect, redundancy is utilized to provide cross-reference listings, DFDs, hierarchical modular models, flowcharts, and alias management. An input-output view of a typical data dictionary package is shown in Fig. 8.9.

---

*A listing of some of the more effective automated data dictionary packages follows: *Data Catalogue* (Synergetics Corporation, Burlington, Mass.); *Data Manager* (Management Systems and Programming, Cambridge, Mass.); *DD/DC* (IBM SB21-1256); *ISDOS* (University of Michigan — Ann Arbor); *LEXICON* (Arthur Anderson, Chicago); *UCC TEN* (University Computing, University of Texas, Dallas).

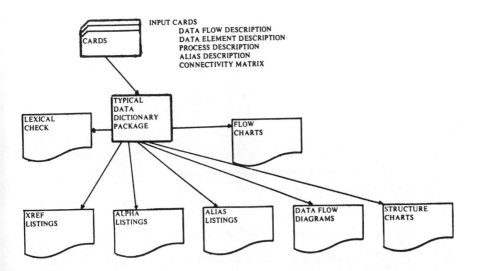

Figure 8.9. I/O view of a typical data dictionary package.

Unfortunately, automated data dictionary packages may be too all-encompassing and expensive for small or medium-size projects; they also may do only part of the job required for large projects. For example, some of these packages document data elements, but not processes within a system. Some produce flowcharts, but not DFDs or structure charts. Some do not facilitate the generation of test data. Others work only with IMS or the equipment of a particular hardware vendor. Most are oriented to documenting the implementation of a system, but are not particularly helpful in the analysis phase of a project. And, almost all of these packages involve a complicated interface that must be incorporated into an organization's procedures.

In summary, automated data dictionary packages manifest both the advantages and disadvantages of generalized products. They have been developed in response to recognized needs; and while most respond to these needs to some degree and aid in controlling the documentation process, none is ideal or the perfect answer for an organization.

Review Exercise

1. Describe a situation in which a misunderstanding of terms can lead to an unclear definition of the required action.

2. Describe at least two data processing situations to show how rushing to a solution leads to problems.

3. Access a program or systems documentation package. Evaluate it. Are all terms and processes properly defined? Could you maintain the system, using the documentation as the reference guide?

4. What are the advantages of documenting a system as it is being developed, as opposed to after the fact?

5. Identify the objectives of a data dictionary. For each objective identified, describe a situation that would result in a problem if that objective were not met.

6. Give three examples to show how a data dictionary with a cross-reference capability could aid in systems development and maintenance efforts.

7. Using Figs. 8.2 through 8.8 as guidelines, define:

   A. (8.2) a data flow named TRANSACTION that could be either an ADD, a DELETE, or a CHANGE

   B. (8.3) a data element named CUST-TELEPHONE-NO

   C. (8.4) a process named UPDATE-MASTER

   D. (8.5) a file named CUSTOMER

   E. (8.6) an alias named ACCOUNT

   F. (8.7) a data element named AMT-OWED

   G. (8.8) a data entry named TIME-SHEET

8. How can a librarian aid in systems documentation?

9. Access the documentation of an automated data dictionary package. Evaluate it. Create a report that depicts both its positive and negative features.

References

1. James Martin, *Principles of Data Base Management* (Englewood Cliffs, N.J.: Prentice-Hall, 1976).

2. R.M. Graham, G.J. Clancy, Jr., and D.B. DeVaney, "A Software Design and Evaluation System," *Communications of the ACM,* Vol. 16, No. 2 (1973), pp. 110-16.

PART 3

Problems in Analysis

9 Strategies to Develop Designs

The purpose of thinking of modules as black boxes is not to defer potential problems, but rather to focus on the proper levels of problems at appropriate times.

9.1 Design perspective

Because most EDP costs are related directly to the testing and maintenance phases of a project, systems architects should consider the overall lifetime costs of systems when creating designs. Historical data indicate that testing consumes close to fifty percent of the total developmental costs of a system, and that maintenance and modifications to the system usually are at least as expensive as the total developmental effort. In some organizations, maintenance and modification consume up to ninety percent of the entire EDP budget. Past studies also indicate that the average system lasts about five years.[1]

Figure 9.1 shows the cost breakdown for the lifetime of a typical system requiring one year and costing $500,000 to develop. During the first year following the system's acceptance, there are substantial maintenance and modifications as the system is thoroughly debugged and enhanced. In the second and third years, costs diminish as the system is stabilized. As time goes on, however, changes in the business environment create a new set of problems requiring system's modifications, and the flexibility of the system is severely tested. Costs increase as the system takes on new functions and revisions. After four years of maintenance, the system, which is held together by switches and coding tricks, usually is so delicate that programmers are afraid to touch it. Modifications and maintenance become very difficult

169

and expensive. Finally, someone says, "It's time to redesign the system," and the developmental cycle begins again.

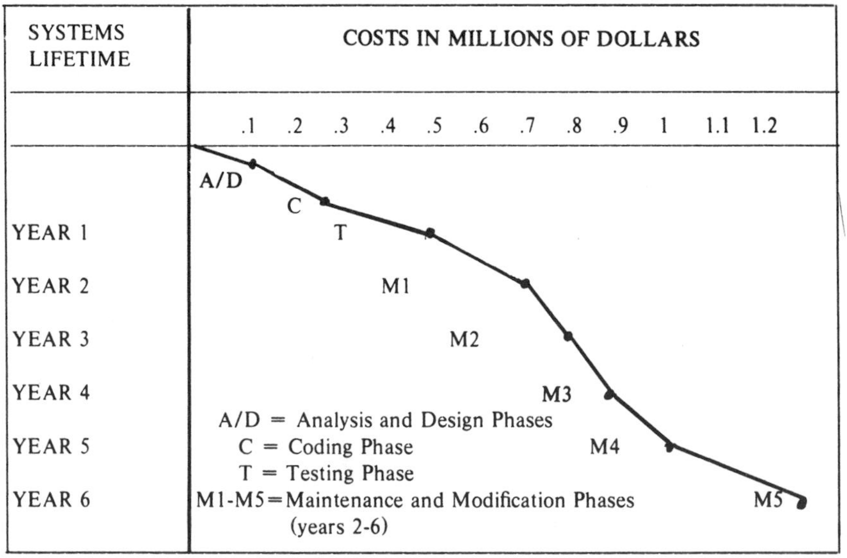

SYSTEMS LIFETIME	COSTS IN MILLIONS OF DOLLARS											
	.1	.2	.3	.4	.5	.6	.7	.8	.9	1	1.1	1.2
	A/D											
YEAR 1		C	T									
YEAR 2			M1									
YEAR 3				M2								
YEAR 4						M3						
YEAR 5							M4					
YEAR 6											M5	

A/D = Analysis and Design Phases
C = Coding Phase
T = Testing Phase
M1-M5 = Maintenance and Modification Phases (years 2-6)

Figure 9.1. Lifetime costs of a typical system.

Some of the reasons why testing, maintenance, and modifications cost so much are the following:

- Program bugs may be hard to find. (The bugs turn out to be somewhere other than where they are expected.)

- Program bugs may be difficult to correct once found. (The code may be tricky or complicated, or the function of the module or routine may not be clear.)

- The correction of one program bug may introduce another. (The code may be interconnected with other functions and other modules.)

- External documentation may be inadequate, with many program listings not commented properly. (A maintenance programmer may be modifying code that he did not write.)

- After years of maintenance, the structure of the logic may be excessively complicated. (Original comments no longer may be relevant.)

- A relatively simple change may involve revising, testing, and documenting many different modules. (Pieces of functions are scattered throughout the system.)

At first glance, these problems may seem to be only programming-oriented, because they inevitably affect programmers. The source of these problems, however, reverts back to the analysis and design phases of systems development, because most of these problems can be avoided with a well-documented, functionally independent, modular design. Unfortunately, most systems analysts and designers have not been trained in design development and evaluation. Moreover, they have been hampered by inadequate tools, vague strategies, and their own enthusiasm. Eager to dive into the specifics of a solution, they frequently do not devote enough time and energy to defining problems, the current environment, systems objectives, or the logical design of the proposed system.

Whenever a proposed system is an expansion, automation, or optimization of a current system, its design and implementation should occur only after the current physical and logical systems and the proposed logical system have been documented. As we saw in Chapter 4, logical design should include data flow diagrams. Logical processes required to transform inputs to outputs* should be defined rigorously in multi-level data flow diagrams, so that bottom-level processes describe logically independent functions. All terms relating to data flows, data bases, and processes of the proposed logical design should be precisely documented in a data dictionary.

---

*Other design strategies, notably those of Jackson,[2] Warnier,[3] Parnas,[4] and Orr,[5] call for the utilization of different design strategies and different communication tools.

After completion of the logical design, systems architects should create hierarchical modular systems. Modules in the physical design should relate to logical processes in the data flow diagram of the proposed system. At or near the top of the hierarchy are "executive" modules, which make decisions about global matters. Detailed "worker" modules, which make fewer decisions and deal with details or only aspects of higher-level decisions, should be located at the middle or bottom of the hierarchy. Figure 9.2 illustrates an abstract hierarchy of modules that indicates a distribution of decision-processing characteristics of well-designed systems.

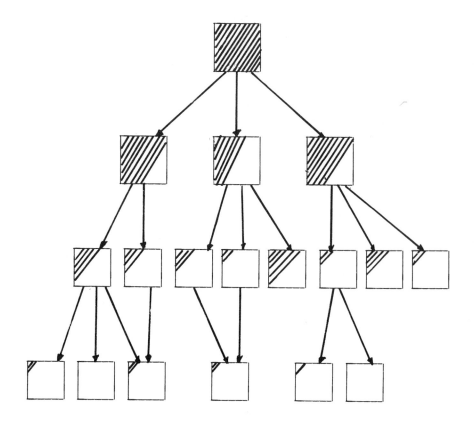

Figure 9.2. Distribution of decision-processing in a modular hierarchy.
(Ratio of decision-processing in modules is shown by diagonal lines.)

In developing a structure chart, systems architects should view each of the modules as a black box. Precise definitions of the inputs, outputs, and the function of each module should be developed now, but detailed definitions of *how* each function will be accomplished can be deferred until later in the physical design process. The purpose of thinking of modules as black boxes is not to defer potential problems, but rather to focus on the proper levels of problems at appropriate times. For example, to know *how* modules will edit, compute, and print it is not as critical early in the design process as it is to recognize that these functions will be required and should be easily maintainable. Modules should be broken down until analyst/designers can envision that the coding required will be manageably small and relatively independent. This process will minimize the numerous problems that can occur when module logic turns out to be much more complicated than originally anticipated.

Another important factor associated with the black-box concept of modules relates to characteristics that will lead to low-cost maintenance: Each module should have only one entry and only one exit point; module interfaces should be designed such that each module in the system has only the data it needs to do its job; and, except in extreme circumstances, modules should not refer to data or to program labels defined within the boundaries of other modules. Maintenance cost is impacted favorably by modules that are functionally independent, manageably small, and correctable separately.

Once a black-box modular design has been completed, systems architects should consider major interfaces in the system, anticipate potentially serious design problem areas, and determine users' priorities in systems implementation. These considerations will help designers to focus on the next design step: creation of detailed specifications for the most important modules in the system. The most important modules are not necessarily the ones in which the most detailed processing will occur, but rather those which, when tested, will prove to analysts, designers, and users alike that the design of the system is sound.

After these modules have been tested and accepted, designers then should develop detailed specifications for the other, less important modules in the system.*

9.2 Top-down design

Top-down design is a general strategy that specifies creating a system's design in terms of its functions.[6] Major functions are defined and then broken down into intermediate functions, which are broken down into detailed, lesser functions, and so on, until functions are sufficiently trivial to be implemented by a manageably small amount of code. Top-down design has the advantage of forcing the designer to consider the major functions (the most important modules) first and the less important ones later. It also forces the designer to consider the amount and nature of the code necessary to implement the design. This general strategy, however, does not specify how to distinguish good designs from bad designs, nor does it provide the analyst or designer with precise guidelines for developing a hierarchical structure of functions. Design strategies that do provide these more specific design guidelines are transform analysis and transaction analysis,[7] discussed below.

9.3 Transform analysis

Transform analysis, or transform-centered design, is a modular design strategy that specifies building a system around the concept of data transformation. The top-level module calls for highly processed logical input data, and lower-level modules transform physical input into logical input, and logical output into physical output. One of the advantages of the transform analysis strategy is that it usually produces systems designs that are easy to develop and to maintain. Another advantage, perhaps more significant, is that it provides a systematic, teachable approach for the analyst/designer; by contrast, top-down design relies heavily on the analyst/designer's intuition.

---

*Other design/implementation approaches are discussed in Chapter 10.

9.3.1 Guidelines to develop transform-centered designs

Using the transform analysis strategy produces a transform-centered structure, as shown in Fig. 9.3, below.

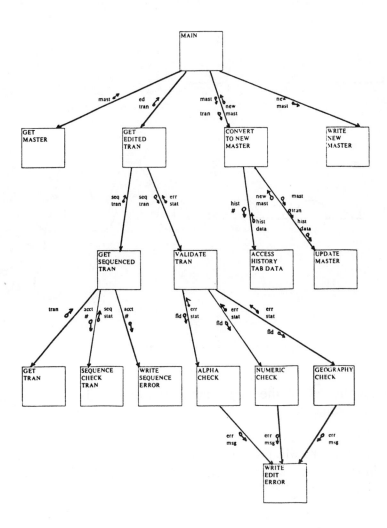

Figure 9.3. Typical transform-centered hierarchical structure.

Steps to develop transform-centered designs are listed on the following pages and are illustrated in Figs. 9.4 through 9.8.

1. Draw a data flow diagram to picture the functions and data transformations in the system. (Refer to Chapter 4 for specific details.) See Fig. 9.4.

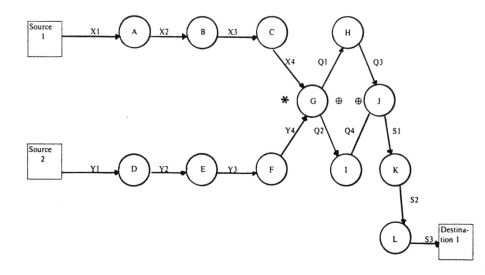

Figure 9.4. Data flow diagram.

2. Identify all of the major input and output data streams in the problem.

3. Follow each input data stream until it has reached its most abstract, highly processed form or until the stream can no longer be considered input. Identify each point; refer to Fig. 9.5 on the following page.

4. Trace each output data stream backward until it can no longer be considered output. Identify each point; refer again to Fig. 9.5.

5. Identify the transformation bubbles (known as "central transforms") in the middle. They process data not recognizable as either input or output. See Fig. 9.5.

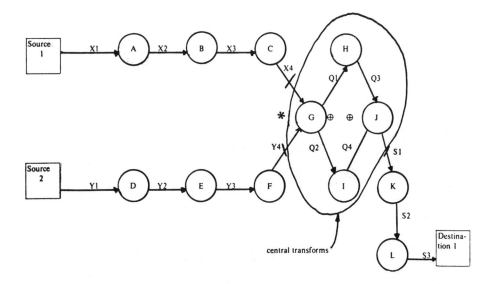

Figure 9.5. Data flow diagram with central transforms identified.

6. Draw the top two levels of a structure chart so that the top module calls one high-level module for each major input stream, one high-level module for each major output stream, and one high-level module for the central transforms. See Fig. 9.6.

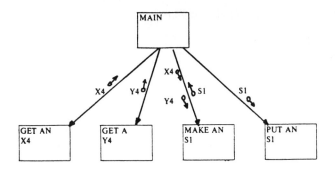

Figure 9.6. Top two levels of a structure chart.

7. Factor the second-level input, output, and central transform modules into their subordinates. See Fig. 9.7.

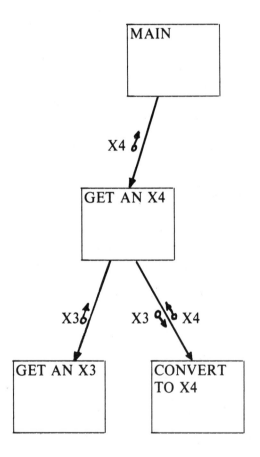

Figure 9.7. Example of second-level factoring.

8. Continue factoring until the entire problem has been depicted in the structure chart (see Fig. 9.8). The lowest modules in the structure chart correspond to the bubbles at the extremities of the data flow diagram, reflecting input-output logic, error logic, and the details of the central transform processing.

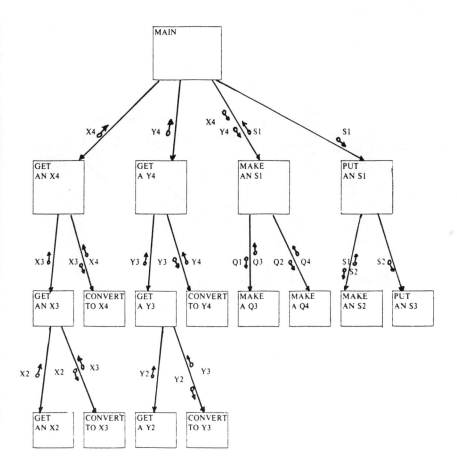

Figure 9.8. Structure chart for entire data flow diagram.

9.4 Transaction analysis

Transaction analysis is another modular design strategy that suggests building a system around the concept of a transaction — that is, any element of data that triggers an action or sequence of actions. Transaction analysis is strongly suggested by data flow diagrams which fan out to do processing by transaction type, as shown by Fig. 9.9. Figure 9.10 depicts a typical transaction-centered structure chart.

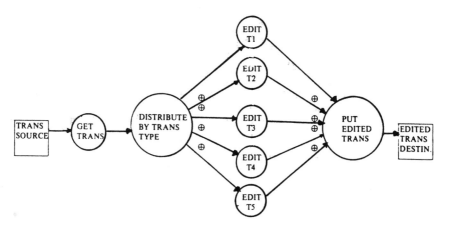

Figure 9.9. Data flow diagram with transaction fan-out.

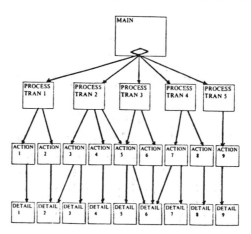

Figure 9.10. Typical transaction-centered modular hierarchy.

9.4.1 Guidelines to develop transaction-centered designs

Transaction analysis produces transaction-centered designs. Guidelines to develop them are listed below:

1. Draw a data flow diagram to picture the functions and data transformations in the system. (DFDs are described in detail in Chapter 4). See Fig. 9.11.

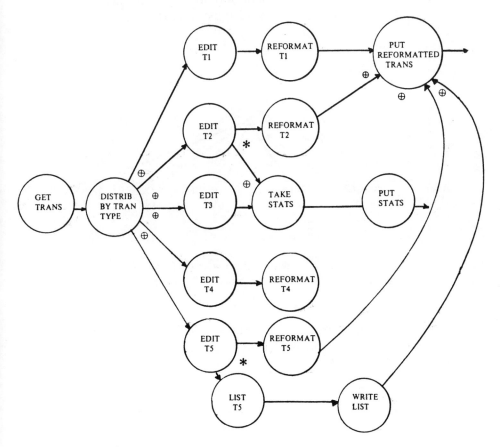

Figure 9.11. Data flow diagram with transaction center.

2. Identify any transformations (bubbles) in which one input data stream produces several mutually exclusive data streams by transaction type.

3.   Identify the transactions and their defining actions.

4.   Note potential situations in which there may be common functions, based upon similarity of transaction composition or syntax.

5.   Draw a structure chart noting the module in which the input data stream is split into several data streams by transaction type. (Use the diamond notation to show the major logic decision.) Figure 9.12 illustrates this.

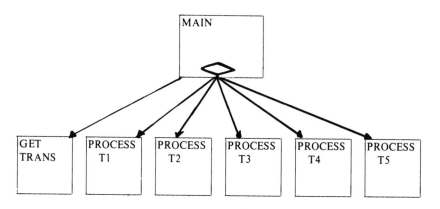

Figure 9.12. Structure chart with transaction center and transaction modules identified.

6.   For each transaction or cohesive collection of transactions, specify a TRANSACTION-module to completely process it. See Fig. 9.12.

7.   For each action in a transaction, specify an ACTION-module subordinate to the appropriate TRANSACTION-module. See Fig. 9.13 on the following page.

8.   For each detailed step in an action, specify a DETAIL-module subordinate to the appropriate ACTION-module. See Fig. 9.13.

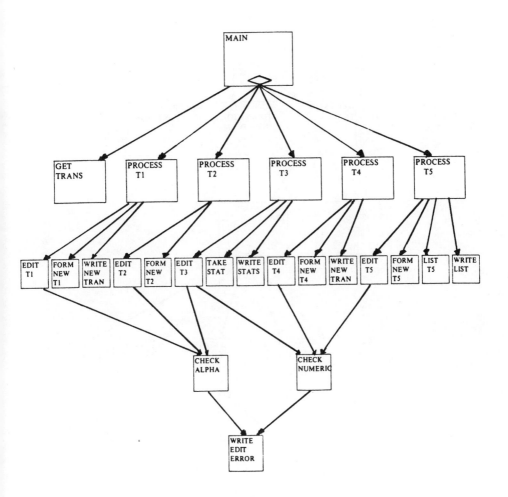

Figure 9.13. Structure chart showing transaction, action, and detail levels.

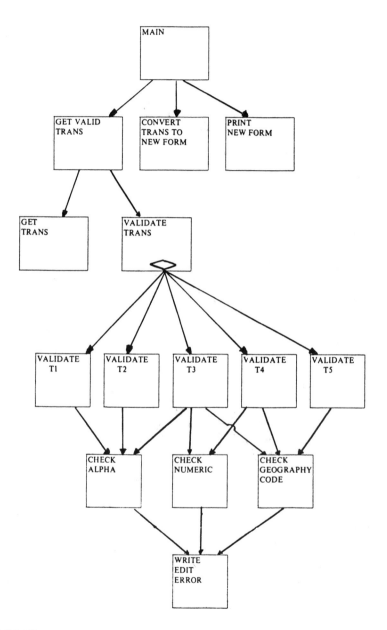

Figure 9.14. Structure chart showing both transform- and transaction-centered desi

Transaction analysis and transform analysis are distinctly different modular design strategies, both of which can be applied to the same problem. For example, Fig. 9.14 pictures a system with unique processing by transaction type as well as a high-level, transform-centered structure. Both strategies start with a data flow diagram and end with a structure chart. The resulting structure chart for this system provides a strong foundation from which subsequent design revision can be made judiciously. Keep in mind, however, that the derived structure chart should not be considered the final modular design, but rather the first *draft* of the modular design. Design evaluation criteria discussed in the following sections should be applied to the modular design to identify potential weaknesses in the hierarchical structure.

9.5 Coupling

Coupling is the measure of strength of association or interconnection between modules.[8,9] Applying this concept helps to assure that a system is comprised of modules that are loosely coupled, that is, relatively independent. Loosely coupled systems reduce potential testing and debugging costs, because the programmer can debug one module without having to know anything about the contents or functions of other modules. Obviously, the more independent a modular structure is, the less likely it is that a change to one module will necessitate modification to other modules as well.

One factor that seriously affects the coupling of a modular system is the complexity of the interfaces between modules. Keeping in mind that "simple" usually is equivalent to "efficient," analysts should assure that interfaces between modules are as simple and as clear as possible. Passing a few parameters is better than passing many. Similarly, if there is a standard way to pass information between modules, the analyst should employ that method. The interface should include the actual data required, when it is available, rather than a pointer to the data. Above all, the interfaces should be obvious, rather than obscure.

In a real-world modular system, none of the modules is totally independent; each must communicate with other modules in the system to transform input data to output data. The *type* of information passed between modules affects the degree of coupling. Frequently, the passing of data between modules is inevitable; less desirable is the passing of control information, such as the passing of switch settings. Although this sometimes cannot be avoided (at end-of-file, for example), designers and analysts both should assure that the system's design does not include excessive switches.

Common programming languages provide the capability to define data in a global way. The DATA DIVISION in a COBOL program, and global variables in a PL/I program are examples of how data can be defined so that it can be accessed by any module in the program. These capabilities should not be overused because they tend to make systems more difficult to maintain. As an example, let's say a program aborts and produces a dump, which shows that a globally defined data field has been mutilated. The question is, How did it happen? The maintenance programmer may have a serious problem answering this question, because *any* module in the program could have destroyed the data. To avoid this problem, we should restrict global access to data to high-level modules. Data should be made available to low-level modules only on a need-to-know basis.

Another factor affecting coupling in a system is the binding time of connections. Binding refers to the process of assigning a fixed value to a piece of information that *could* change; it can occur at execution time, loading time, linkage-edit time, or, least desirable, at the instant the programmer writes the code on a coding sheet. This last situation usually is known as "coding with literal constants." To understand how binding affects systems coupling, consider the nature of changes that would be required to a system if the number of entries in a table changed from 15 to 16. The following points illustrate this:

- If the programmer hand-coded literals throughout the system whenever the table was accessed, each one of those literals would have to be changed. The effort, which may require numerous recompilations, is prone to error in that literals are not always cross-referenced in program listings.

- If the programmer created a data element called TABLE-LIMIT to be used instead of literals whenever the table was accessed, only one program statement would have to be changed and one module recompilation would result.

- If the number of table searches was a parameter on card input, only the card input would change. In effect, the program or system would require no modification or recompilations.

The evaluation of coupling sometimes is not as easy as it may appear. For example, Fig. 9.15a shows that editing may result in an error, which will be printed. The EDIT modules, upon finding an error, format an ERROR LINE and pass it to the calling module. The calling module, GET EDITED XACT, passes the ERROR LINE to a module, which prints the error. Figure 9.15b shows a different solution to the same problem. In this case, an error is handled immediately by a direct call to the PRINT ERROR module. The specific data associated with the error is screened from the GET EDIT XACT module in that it sees only an ERROR STATUS indicator. The question is, Which modular structure, shown in our two solutions, has less coupling?

At first glance, Fig. 9.15b seems more complicated than Fig. 9.15a in that it requires an additional level in the modular hierarchy and an ERROR STATUS switch. The number of levels in a hierarchy, however, bears no relationship to systems coupling, and the ERROR STATUS switch, notably absent in Fig. 9.15a, is not really absent. In Fig. 9.15a, ERROR LINE is used both as data and control. The GET EDITED XACT module has to look closely at ERROR LINE to determine whether it is null or contains an error message; in this case, ERROR LINE is both a switch and an element of data. This analysis leads to the conclusion that Fig. 9.15b has less coupling than Fig. 9.15a, because Fig. 9.15b requires the passing of fewer parameters. This analysis is depicted in Table 9.1.

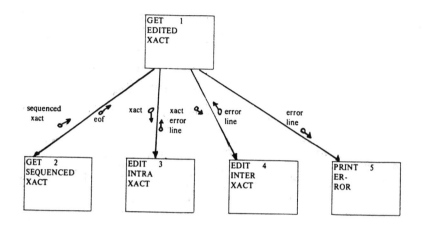

Figure 9.15a. Hierarchy coupling evaluation — possible solution.

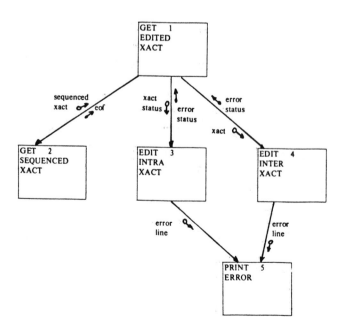

Figure 9.15b. Hierarchy coupling evaulation — another solution.

Table 9.1
Coupling and Hierarchy Parameter Analysis

Figure 9.15a	Input Parameters	Output Parameters	Total Parameters
GET SEQUENCED XACT		SEQUENCED XACT EOF	
EDIT INTRA XACT	XACT	ERROR LINE (ERROR STATUS)	
EDIT INTER XACT	XACT	ERROR LINE (ERROR STATUS)	
PRINT ERROR	ERROR LINE		9
Figure 9.15b			
GET SEQUENCED XACT		SEQUENCED XACT EOF	
EDIT INTRA XACT	XACT	ERROR STATUS	
EDIT INTER XACT	XACT	ERROR STATUS	
PRINT ERROR	ERROR LINE		7

9.6 Cohesion

Cohesion is a measure of the degree to which the elements within a given module relate to the accomplishment of a simple, identifiable task. Ideally, all statements or elements within a module should be strongly interrelated, that is, highly cohesive. Table 9.2 shows the various levels of cohesion.[10,11] When evaluating the cohesion of a module, the analyst or designer may find that a module belongs in more than one of the categories outlined in Table 9.2. In this event, the module is said to have the highest applicable level of cohesion.

For the most part, systems comprised of independent, functionally cohesive modules are the most desirable. Sometimes, however, two or three functions may involve so little coding that the functions are combined into one module. In other situations, we may want to group certain independent functions together because they all are time-related (initialization, etc.), or they all involve one general task. In another instance, we may want to repackage the system by combining modules, perhaps because the implementation of the system is too slow or takes up too much space. Generally, however, the modules in the system sl ꞈuld be functionally independent. In those modules exhibiting other forms of cohesion, creation of switches and tricky code to distinguish between various module subfunctions should be minimized.

Table 9.2
Categories and Levels of Cohesion

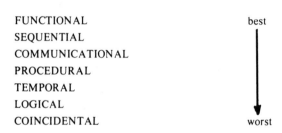

FUNCTIONAL	best
SEQUENTIAL	
COMMUNICATIONAL	
PROCEDURAL	
TEMPORAL	
LOGICAL	
COINCIDENTAL	worst

A module with *coincidental cohesion* contains elements that have no meaningful relationship to one another. Coincidental cohesion should be avoided because it results in severe debugging problems. When a bug occurs or a change has to be made in a system with coincidentally cohesive modules, the programmer may have to analyze the contents of almost every module in the system in order to solve the problem.

Coincidental cohesion usually results from laziness on the part of the programmer. To illustrate, assume that a new standard has been developed requiring no module to be larger than one hundred statements. A programmer, who has just finished coding a module that contains one thousand statements, decides to adhere to the letter, but not the spirit, of the law. He breaks the module into ten modules, each with one hundred statements in the order of the original thousand-statement module. When viewed separately, these ten new modules do not seem to have a high level of cohesion. To the contrary, they perform only subfunctions and, in many cases, even subfunctions are split between modules. These modules have coincidental cohesion in that their physical boundaries are the results of the hundred-statement programming standard, rather than the functions required in the system.

A module with *logical cohesion* contains elements that all are logically related in performing similar tasks. In most cases, a logically cohesive module performs general-purpose processing, such as a general-purpose input or error routine. Logically cohesive modules sometimes are desirable in systems that have potential space problems, because the general nature of the functions of these modules can lead conveniently to shared constants, buffers, and code.

Unfortunately, the savings in space frequently is accomplished with negative side effects. The sharing of constants and code seriously can complicate the debugging of a system. In addition, switches and program conditions must be tested to distinguish which of the many potential subfunctions is being invoked on each call to a logically cohesive module; as a result, logically cohesive modules tend to slow processing. Perhaps the most serious problem is that changes to logically cohesive modules are not easily made because of the shared constants and coding; the system's consequent inflexibility results in unnecessarily high maintenance costs. For these reasons, logically cohesive modules generally should be avoided.

A module with *temporal cohesion* contains elements all related in time. Examples of modules with temporal cohesion are initialization, termination, and housekeeping routines. Temporally cohesive modules are preferable to logically cohesive modules, because the logic is less complicated; there are fewer switches; all elements in the modules are executed; and the elements generally can be executed in any order. Normally, however, temporally cohesive modules combine too many functionally independent tasks. For example, a programmer looking at a module named TERMINATION, which has failed, may have difficulty determining which of the many termination functions is associated with the program failure. Because the module name's does not adequately describe its detailed functions, unnecessarily high maintenance and debugging costs may result. In addition, a temporally cohesive module may contain only fragments of various functions, the other fragments being contained in other modules. This leads to coupling problems. When a programmer has to modify the way in which a system handles disk input, for example, the DISK-READ module and the INIT module (which opens and initializes the disk file) both may have to be changed.

A module with *procedural cohesion* is characterized by control flowing from one element to the next, and usually results from the combining of a few flowchart sections into one module. For example, a procedurally cohesive module might clear work areas, reset switches, accumulate totals, and close files. While considered stronger than previously defined levels of cohesion, procedurally cohesive modules exhibit some undesirable characteristics. One basic problem is that all elements within a module do not necessarily deal with the same data. Another weakness is that procedurally cohesive modules may deal with only one part of an overall function, or even worse, fragments of several functions. When a problem arises relating to a particular function, the maintenance programmer may have difficulty finding the associated pieces.

A module with *communicational cohesion* contains elements that reference the same data. A module that prints or punches a transaction file, a module that both prints and punches a transaction file, and a module that does all input and output processing for a particular file are examples of communicationally cohesive modules. A communicationally cohesive module is much like a general-purpose module in that it may do many functions, switches may be created to distinguish functions, and all elements may not be executed. Communicationally cohesive modules are more desirable than previously defined levels of cohesion. This is because of the continuity of the data between the elements.

A module possessing *sequential cohesion* exists when the output data from one element becomes the input data to the next element. Examples are a module that edits a transaction and then updates a master record with the edited transaction; or a module that updates a record and then prints the updated record. Note that, in the first case, the edit must precede the update whereas, in the second case, the update must precede the print. In a module that both prints and punches a record (communicational cohesion), the activities can be done in any order. Sequentially cohesive modules, by contrast, require a particular sequence of actions and a continuity of data among all the actions. Sequentially cohesive modules, however, may be involved in doing only parts of a function or multiple functions. The nonfunctional aspects of these modules may result in debugging and maintenance problems.

A module with *functional cohesion* contains elements that all contribute to the execution of only one function. Examples of functionally cohesive modules are a module that computes elapsed time, one that edits a payroll transaction, a module that prints a daily report, or one that reads an inventory record. Note that these examples do neither more nor less than what has been stated. Functionally cohesive modules are relatively easy to debug and maintain because their functions are clear, and their logic is straightforward.

```
IF SENTENCE IS A COMPOUND SENTENCE,
         OR CONTAINS A COMMA,
         OR CONTAINS MORE THAN ONE VERB,
      THEN THE MODULE PROBABLY HAS COMMUNICATIONAL,
      SEQUENTIAL, OR LOGICAL COHESION
ELSE
IF SENTENCE CONTAINS TIME-ORIENTED WORDS SUCH AS
         FIRST, LAST, NEXT, AFTER, START, INITIALIZE,
         CLEANUP, TERMINATE, ETC.,
      THEN THE MODULE PROBABLY HAS SEQUENTIAL, PROCEDURAL,
      OR TEMPORAL COHESION
ELSE
IF SENTENCE CONTAINS A PLURAL OR COLLECTIVE OBJECT,
      THEN THE MODULE PROBABLY HAS LOGICAL OR
      COMMUNICATIONAL COHESION
ELSE
IF SENTENCE MAKES NO SENSE OR CONTAINS NO CONTINUITY OF
         FUNCTION,
      THEN THE MODULE PROBABLY HAS COINCIDENTAL COHESION
ELSE (NONE OF THE ABOVE)
      THEN THE MODULE HAS FUNCTIONAL COHESION.
```

Figure 9.16. Guidelines to identify categories of cohesion.

As stated earlier, a module may exhibit more than one level of cohesion. One reasonably reliable way of determining the highest applicable level of cohesion is to write a sentence describing accurately and completely the purpose of the module. Then apply the guidelines expressed in structured English in Fig. 9.16 to identify the applicable category or categories of cohesion.

These guidelines are useful so long as the description expresses what the module is doing, not how the module is doing it. For example, consider a module named COMPUTE-CUMULATIVE-GROSS-PAY, which does nothing but that; it is clearly a functionally cohesive module. If, however, the sentence describing that module contained the detailed steps to achieve its function, then the application of these guidelines would be misleading. See if you can determine the level of cohesion from the following description:

First multiply the number of hours worked by the pay rate, store the result in gross pay, and then add the gross pay to the cumulative gross pay, storing the result in the cumulative gross pay.

In that the above sentence contains commas, more than one verb, time-oriented words (first, then), and involves a necessary sequence of events with a continuity of data between steps, we might identify temporal, procedural, communicational, and sequential cohesion. We should keep in mind, however, that all of the verbs (multiply, add, store) describe detailed, instruction-level actions and that these detailed steps contribute to a single function: to compute the cumulative gross pay. As a result, the module has functional cohesion, regardless of the way that function is described.

What we should be aiming for in systems design is module predictability and, consequently, module maintainability. A module's name may suggest functional cohesiveness, but frequently the module may be doing more than one function. For example, the module EDIT may print an error line when the transaction it is editing is in error. The printing of the error line is considered a side effect of the main function of editing. This side effect, however, may not be apparent from the module's name or from a study of the structure chart of the program.

Imagine the potential problem when a maintenance programmer must find where to modify the coding that prints the error line. If neither the module name nor the structure chart points out the location of the problem, the maintenance programmer may have to open the listing for each module in the system and scan thousands of lines of code to find the appropriate error-handling routine. By changing the name of the EDIT module so that it more properly describes the function of the module (EDITANDERROR), or by creating a separate and independent module (PRINT-TRAN-ERROR) to support the overall edit function, the systems designer can produce modules that are both predictable and maintainable.

9.7 Morphology

Morphology refers to the shape of a structure. In a systems design environment, morphology refers to the shape of the overall hierarchical modular structure. While shape of a system does not determine the quality of a design, well-designed systems tend to fan out at the top and fan in at the bottom of the modular hierarchy, as shown in Fig. 9.17. The fan-out at the top is created by executive or managing modules controlling the flow of the detailed work done by lower-level modules. Fan-in at the bottom of the hierarchy is caused by detailed work modules that are required as subfunctions by various modules in the system.

The number of modules that should be subordinate to an executive or high-level module depends upon the specific requirements of the system. Studies of personnel hierarchies in an organizational structure have suggested that most executives and managers should control not more than nine (seven plus or min.us two) subordinate workers. [12,13] The reasoning behind this guideline is that too few subordinates constitutes too many chiefs and not enough Indians; more than nine subordinates suggests that there may be too many Indians and not enough chiefs. The seven-plus-or-minus-two guideline has proven to be reasonably sound when applied to the modular design of systems.

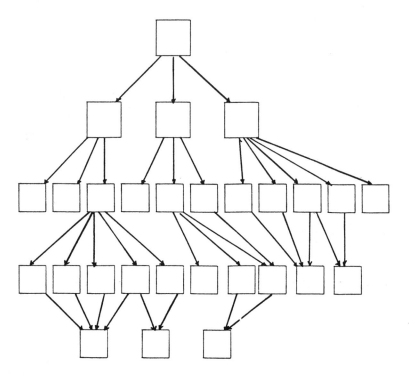

Figure 9.17. Fan-out and fan-in characteristic of well-designed systems.

There are characteristics of systems, however, that suggest situations in which the guideline will not be applicable. At the very top of the hierarchy, as shown by Fig. 9.17, the main executive module frequently will have only a few subordinate modules. At the lowest levels of the hierarchy, where module logic is shared by superordinate or calling modules, the calling modules may be expected to have only one or two subordinates. The only time we would expect a module to have more than nine subordinate modules is in a transaction-centered system. When a system has to edit and/or update with a transaction file comprised of thirty different transactions, normally there will be one module that determines the transaction type or group of transactions. In such a system, we might expect the transaction-identifying module to have more than nine subordinate modules.

9.8 Scope of control/scope of effect

A common weakness in modular designs is the excessive creation and passing of switches. A switch normally is created when some condition is tested, and the results of the test must be remembered so that another part of the system can do its work. One obvious way to eliminate the switch is to retest the condition, if the data still are available. But now, instead of an excessive switch problem, the design exhibits duplicate or triplicate decision-making. Repetitive switch testing and decision-making both are indicative of a poorly designed system, because modular designs with these characteristics are overly complicated and inefficient.

Fortunately, structure charts, showing major decisions and the flow of switches, enable us to identify these problems early in the physical design process. Terminology to define these problems and guidelines or heuristics for their correction have been developed by Constantine and Yourdon,* and have been adapted as follows:

DEFINITIONS:

The scope of control of a module is that module and all modules subordinate to it.

The scope of effect of a decision is all modules affected by that decision.

HEURISTICS:

1. The scope of effect should be within the scope of control.

2. A decision should be made no higher in the hierarchy than is necessary to place the scope of effect within the scope of control.

---

*E. Yourdon and L.L. Constantine, *Structured Design: Fundamentals of a Discipline of Computer Program and Systems Design,* 2nd ed. (New York: YOURDON Press, 1978), Chapter 9.

To further illuminate these scope of control/scope of effect definitions and heuristics, assume that a program is comprised of eight functionally cohesive modules as shown in Fig. 9.18. The scope of control of module A includes all of the modules in the program; the scope of control of module B includes modules B, D, and E; and the scope of control of module C includes modules C, F, G, and H. The diagram also indicates that there is a major program decision made in module B. This decision determines whether module E is called. It also results in a switch, shown as X, which is passed up and down through the program, because some action in module H depends upon its status.

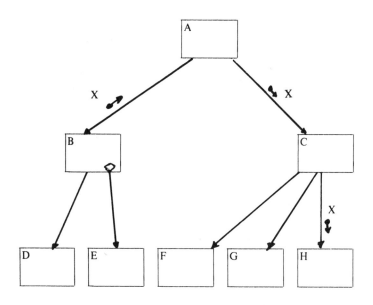

Figure 9.18. Scope of control/scope of effect violation.

Note that Fig. 9.18 shows a violation of the first scope of control/scope of effect heuristic: The scope of effect of a decision should be within the scope of control of the module making the decision. The scope of effect of the decision made in module B includes modules B, E, and H. Module H is not within the scope of control of module B.

Figure 9.19 shows a variation on the same theme. In this case, however, instead of passing a switch up and down the hierarchical modular structure to remember the status of a decision, the decision itself is duplicated as shown by the diamond notations in modules B and H. The scope of effect of the decisions is within the scope of control of the modules making the decisions, but now the program exhibits duplicate decision-making. Duplicate decision-making and excessive switch passing really are the same problem camouflaged by different symptoms.

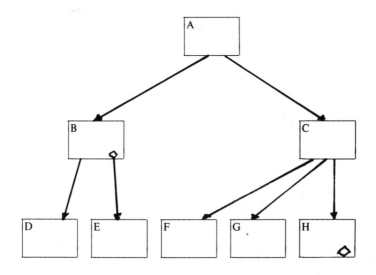

Figure 9.19. Another violation — duplicate decision-making.

The question is how to handle scope of control/scope of effect problems. Essentially, the options are to move the decisions, to rearrange the modules, or to live with the problem. For example, we might remove the scope of control/scope of effect violation shown in Fig. 9.18 by moving the decision from module B to module A, but we may create in its place two new problems. As shown in Fig. 9.20, module B now may be doing only part of an overall function, or module A may be doing more than one function. Furthermore, we have not reduced the amount of switch passing in the system.

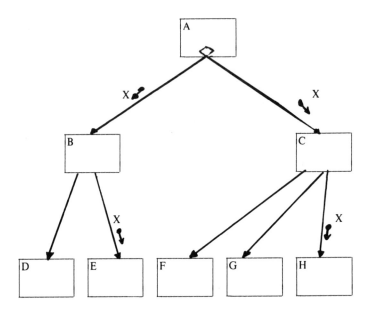

Figure 9.20. Moving the decision.

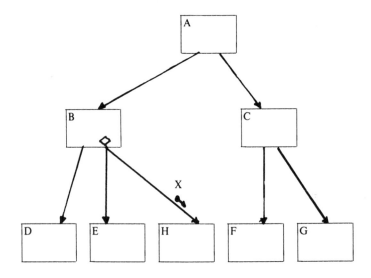

Figure 9.21. Moving a module.

Another approach is to move module H so that it is subordinate to module B as shown in Fig. 9.21. This solution neither violates any heuristic nor erodes the modular cohesion of the program. The solution, however, may not be feasible if the processing in module H must be logically preceded by the processing in modules C, F, and G.

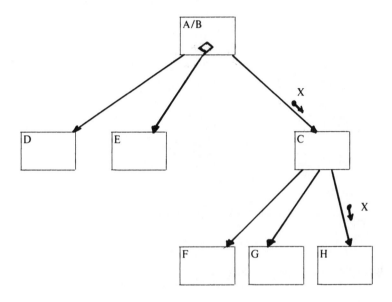

Figure 9.22. Combining modules.

Another option is to combine modules A and B, as shown in Fig. 9.22. This eliminates the upward passing of a switch in the modular structure and results in the scope of effect of the decision being within the scope of control of the module making the decision. Module A/B, however, may be unmanageably complicated in that it now is doing work that normally would be separated into two functional modules.

Still another approach makes all modules directly subordinate to module A as shown in Fig. 9.23. This eliminates the duplicate decision-making and minimizes switch passing. Module A, however, now may be unmanageable because it controls the functions of so many modules.

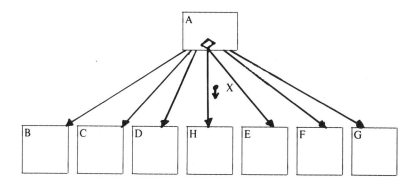

Figure 9.23. Rearranging modules.

9.9 Design evaluation summary

Design strategies such as transform and transaction analysis provide a relatively straightforward approach to produce modular designs from data flow diagrams. These tools and strategies, however, only result in a preliminary hierarchy of modules. Design guidelines such as cohesion, coupling, morphology, and scope of control/scope of effect must be applied next to improve the initial design.

By now, it should be quite apparent to all that these separate design evaluation strategies are interrelated. Solving a scope of control/scope of effect problem may create a coupling, cohesion, or morphology problem. Similarly, solving a cohesion problem may create a coupling problem, and so on. To some extent, design evaluation results in a damned-if-you-do, damned-if-you-don't situation. As soon as you push down one lump in the bed, another lump shows up.

Application of these design evaluation guidelines will not necessarily result in clear-cut answers. For example, since levels of cohesion have not been quantified, we cannot say a module with sequential cohesion is three times as maintenance-free as a module with temporal cohesion (in fact, there are times when a temporally cohesive module might be more desirable than one with a higher level of cohesion). Moreover, no firm guidelines exist to evaluate the interrelationships of various kinds of design problems. While good cohesion is perhaps the most important concept (and a direct result of precise functional decomposition at the data flow diagram level), analysts and designers still find themselves in a gray area when evaluating physical designs. Precise terminology and concepts such as cohesion, coupling, morphology, and scope of control/scope of effect can only help to make design development a more black-and-white process.

Review Exercise

1. Describe the use of the data flow diagram in relation to the systems design process.

2. How does the data dictionary support the systems design process?

3. How does the concept of black boxes relate to the systems design process?

4. How does transform analysis differ from transaction analysis?

5. What are the differences between top-down design and transform or transaction analysis?

6. How can a structure chart aid in the evaluation of systems designs?

7. How are the concepts of coupling, cohesion, morphology, and scope of control/scope of effect interrelated? Show the interrelationships, using at least two examples.

8. What is meant by a side effect of a module and how do module side effects lead to problems in maintenance?

References

1. B. Boehm, "Software Engineering: R and D Trends and Defense Needs," *Proceedings of the Conference on Directions in Software Technology,* 1977.

2. M.A. Jackson, *Principles of Program Design* (New York: Academic Press, 1975).

3. J.D. Warnier, *The Logical Construction of Programs,* 3rd ed., trans. B.M. Flanagan (New York: Van Nostrand Reinhold, 1976).

4. D.L. Parnas, "On the Criteria to Be Used in Decomposing Systems into Modules," *Communications of the ACM,* Vol. 15, No. 12 (December 1972), pp. 1053-1058.

5. K.T. Orr, *Structured Systems Development* (New York: YOURDON Press, 1977).

6. E. Yourdon, *Techniques of Program Structure and Design* (Englewood Cliffs, N.J.: Prentice-Hall, 1975).

7. E. Yourdon and L.L. Constantine, *Structured Design: Fundamentals of a Discipline of Computer Program and Systems Design,* 2nd ed. (New York: YOURDON Press, 1978).

8. Yourdon and Constantine, op. cit., pp. 76-94.

9. G.J. Myers, *Reliable Software Through Composite Design* (New York: Petrocelli/Charter, 1975), pp. 33-53.

10. Yourdon and Constantine, op. cit., pp. 95-126.

11. Myers, op. cit., pp. 19-31.

12. G.A. Miller, "The Magical Number Seven, Plus or Minus Two: Some Limits on Our Capacity for Processing Information," *Psychological Review,* Vol. 63 (1956), pp. 81-97.

13. P.F. Drucker, *Management: Tools, Responsibilities, Practices* (New York: Harper & Row, 1974).

10 Strategies for Systems Implementation

One of the most distressing aspects of large systems development efforts from the users' point of view is that, typically, no tangible results are seen until the project has been completed.

10.1 An old story

Systems implementation strategies seriously affect both the schedule and the costs estimated during prior phases of a developmental project. Too often, immediately after a project implementation team gets the system's design and programming specifications, they start coding the least complex modules. Testing of these modules goes well because the functions being tested are trivial. Then, combining some of the pieces, they discover a few problems. As they correct program bugs, some late changes to the specifications are delivered. Modules are revised; a few new modules are created; and the more that pieces are put together, the more the system seems to fall apart.

As the project continues, program bugs become increasingly difficult and time-consuming to locate and fix. As the deadline approaches, people are added to the project. The few people who really know the system spend their time training the new personnel.[1] Time and tempers grow short and communication problems develop as programmers, analysts, and managers work under severe pressure for six or seven days per week. Finally, the last modules in the system are coded, but nobody has created the appropriate test data.

When the test data do become available, the project team is so pressured that it tests all of the new modules together. Testing results are disastrous. Formats accepted by one part of the system are unacceptable in other parts of the system. New test data reveal bugs in supposedly successfully tested logic. The bugs require redesign and recoding of many modules in the system. The Operations Department complains about the excessive test-time required by the project. The team manager knows that the project has lost control of the testing, but it seems too late to go back to take a more methodical approach. The deadline is not met, and the workers continue to stumble through the testing phase until either they wear down the problems or the problems wear them down. Either the system never gets finished, or chances are good that it will be substantially late and well over the estimated budget.

For all of their efforts, the project team has not met the cost and schedule objectives of the users. The programmers are worn out by what seems to have been a thankless task and the project manager has lost credibility. The only fruits of the team's efforts are ulcers, a lot of gray hair, and perhaps a few termination interviews with the Personnel Department.

The preceding story is a familiar one. From such experiences, we should be able to determine classical weaknesses in implementation strategies and apply techniques to minimize their occurrences. From the above story, for example, we can identify the following errors in judgment:

- Programmers on the team began coding the wrong modules first.

- They did not get to the serious, more difficult bugs until late in the project.

- Untrained people were added late in the project.

- The test-data-creation effort was not coordinated with the coding effort.

- The team tried to test too much at once.

- The project manager accepted changes to the systems specifications without updating the budget and schedule accordingly.

- The team had no plan to give the users part of the system before the entire system was to be delivered.

10.2 The implementation plan

Perhaps all of these problems resulted in part from the project team's enthusiasm to start coding immediately. The larger the effort, the more critical it is that implementation not begin until an overall plan has been developed and approved. The plan should identify all of the activities and responsibilities of the total implementation effort. It should include not only activities and responsibilities for programming, but also those related to any task required to assure a successful systems development effort. (See Tables 10.1 and 10.2.) For example, who is responsible for creating test data and verifying test results? Who must identify and enforce systems standards? Who is to identify training requirements and assure a coordinated training program? Who will prepare operational procedure manuals and verify that these procedures are both correct and understood? Who will coordinate with outside vendors to assure that their products work and are properly documented? And, perhaps most important, who will manage and coordinate the overall systems development effort?

Table 10.1: Departmental Responsibility Chart

ACTIVITY	USER INTERFACES			SYSTEM'S IMPLEMENTORS			
Customer Conversion	Sales	Publishing	Quality Control	Program- mers	Opera- tions	Tech. Analysts	Clerical
1. Conversion/Edit (CUST100)	C,R,A	C,R,A	C,R,A	D	C,R,A	C,M	S
2. Update/Report (CUST200)	C,R,A	C,R,A	C,R,A	D	C,R,A	C,M	S
3. Test Data Creation	D	D	C			C	S
4. Standards Development	C,R	C,R	D,M,R,A	C,R	C,R	C,R	S
5. User Procedures	C,R,A	C,R,A	C,R,A			D,M	S
6. Training	C,R,A	C,R,A	D,M			C	
7. System's Acceptance	R,A	R,A	C,R,A,M			C	

Legend: A = Approve; C = Consult; D = Develop; M = Manage;
R = Review; S = Support.

Table 10.2
Detailed Activity Schedule/Responsibility

DETAILED ACTIVITY 4.0 Standards Development	SCHEDULE IN WEEKS DEADLINE 1 2 3 4 5 6 7 8 9 10 11 12 APRIL　　MAY　 JUNE 1 8 15 22 29 6 13 20 27 3 10 17	DEPT. MANPOWER RESPONSIBILITY
4.1 Identify standards and procedures currently utilized	——	5 days/QC 1 day/Sales; 1 day/Publ; 1 day/Prog; 1 day/Oper
4.2 Determine the adequacy and limitations of current standards and procedures	—	5 days/QC ½ day/Sales; ½ day /Publ; ½ day/Prog; ½ day/Oper
4.3 Develop a draft of proposed standards with enforcement procedures for:		15 days/QC; 10 days/Clerical
• Programming	———	5 days/QC; 3 days/Clerical
• Testing	—	5 days/QC; 3 days/Clerical
• System's Acceptance	—	5 days/QC; 4 days/Clerical
4.4 Distribute separate drafts	– ––	½ day/Clerical
4.5 Review and evaluate standards	————	discretionary by dept.
4.6 Revise standards until accepted	————	maximum of 10 days/QC 5 days/Clerical; 2 days/Sales; 2 days/Publ; 2 days/Oper

Quality Control:	J. Smith (Standards Coordinator)	Publishing:	M. Brady (Director)
Programming:	A. Jackson (CUST100) C. Brown (CUST200)	Operations:	P. Kelly (Day Shift Supervisor)
Sales:	B. Jones (Marketing Manager)	Clerical:	D. Wilson (Technical Secretary)

The implementation plan also should detail project cost estimates and schedules. Logical processes are broken down to automated and manual activities, which are assigned to specific personnel with known experience and expected productivity capabilities. Large projects are broken down into sub-projects or phases of development, which can be more easily managed and measured. In general, as project activities and responsibilities are refined, so are estimates of project costs and schedules.

One of the most distressing aspects of large systems development efforts from the users' point of view is that, typically, no tangible results are seen until the project has been completed. By that time, the money has been spent; if the users do not like the product, the damage already has been done. Even if the product can be changed, revision is a costly and time-consuming effort. Part of the problem is that the users may not know whether they will like a product until they start to use it. For this reason, the implementation plan should state implementation objectives clearly and should specify delivery of preliminary versions. In this way, users can work with and evaluate a subset of the system long before the entire system has been implemented and delivered.

We have seen some of the problems caused by the lack of an effective implementation plan. Now let's consider the steps to be taken: The implementation plan, like many other phases of a systems development effort, should be thought of as an iterative process. Systems designers, analysts, and users first should get together to define a general plan for the entire systems implementation, including definition of deliverables and refined estimates relating to schedules and costs. Then, they should determine how to treat the overall effort so that the least amount of time and money will test the validity of the design and produce some tangible output, which can be evaluated by the users. Once intermediate versions or releases of the system are defined, a detailed implementation plan should be developed for preliminary versions. As preliminary versions of the system are implemented and accepted, detailed implementation plans should be developed for subsequent versions until the final version of the system has been implemented and accepted.

The coordination of testing is a particularly serious problem to be addressed in the implementation plan. Many project managers have watched implementation deadlines slip by because of a loss of control in testing. We saw in our story at the beginning of the chapter that programmers tend to implement the wrong modules at the wrong time. Preferring to focus on details rather than on abstractions, they work bottom-up, implementing low-level, detailed modules first. As a result, they need to create special test data and "driver" programs to simulate control logic; these typically are discarded as soon as the low-level modules have been tested. As the modules are linked together, major interface problems and problems with low-level modules are encountered.

Many of these problems can be eliminated by taking a top-down, rather than bottom-up, approach to testing. If top-level modules are implemented before lower-level modules, the need for driver programs is eliminated and major interface problems are exposed, before they affect the logic of lower-level modules. If tests are designed so that each successive test retests already accepted logic, each version of the system will provide the users with more functional capability, and logic accepted still will be valid in later versions. And, if combinations of modules are tested incrementally (by linking a manageably small module or group of modules to a working subset of the system), better control of testing will result and systems and integration testing virtually will be eliminated as separate steps.

Incremental testing is an essential aspect of top-down testing philosophy. Instead of unit-testing fifty modules and linking them together in one test (only to discover that the system does not work and the error is hard to find), systems should be tested in a more controlled fashion. A small, high-level subset of the system should be tested until it works. Since this high-level subset calls lower-level modules, lower-level modules should be simulated. These simulated modules or "stubs" contain real module linkage logic, but do not actually do all of the lower-level detailed work. A module stub, for example, could be a primitive or incomplete version of a real module; it could return a message or constant output, or simulate the timing of lower-level work before exiting. In any case, once a high-level subset of the system is working, testing is controlled by attaching a manageably small amount of untested code to the working subset of modules until the new, larger subset of the system is acceptable. This process is continued until the last module or modules of the system have been added to the ever-growing subset of tested modules.

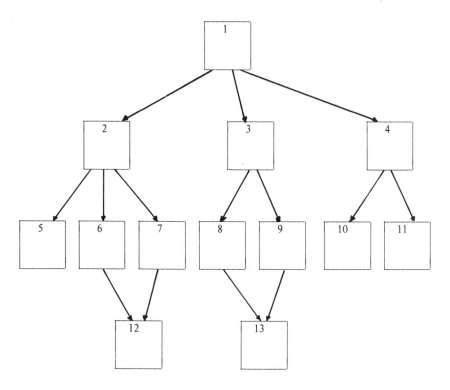

Figure 10.1. A program with thirteen modules.

The implementation plan should indicate clearly the specific sequence of tests anticipated for every version of the system. As an example, let us assume a program is made up of 13 modules (or groups of modules), as shown in Fig. 10.1. Each module in the program is numbered to indicate the sequence in modular coding and program testing. One can see that if module 12 were to be coded before modules 1 and 2, module 12 could not be tested without a driver program.

A modular test plan should be developed for all module testing in the system. Table 10.3, for example, indicates that the 13-module program will be tested in 13 separate tests. The table indicates the status of all program modules required for each phase of testing. It clearly indicates which modules are required as stubs, which are real, and which are not needed for any stage in testing. Project managers have found this technique to be of considerable value in controlling testing and in measuring progress of implementation.

Table 10.3
Modular Test Plan

TEST	MODULE STATUS												
	1	2	3	4	5	6	7	8	9	10	11	12	13
1	R	S	S	S									
2	R	R	S	S	S	S	S						
3	R	R	R	S	S	S	S	S	S				
4	R	R	R	R	S	S	S	S	S	S	S		
5	R	R	R	R	R	S	S	S	S	S	S		
6	R	R	R	R	R	R	S	S	S	S	S	S	
7	R	R	R	R	R	R	R	S	S	S	S	S	S
8	R	R	R	R	R	R	R	R	S	S	S	S	S
9	R	R	R	R	R	R	R	R	R	S	S	S	S
10	R	R	R	R	R	R	R	R	R	R	S	S	S
11	R	R	R	R	R	R	R	R	R	R	R	S	S
12	R	R	R	R	R	R	R	R	R	R	R	R	S
13	R	R	R	R	R	R	R	R	R	R	R	R	R

Legend: R = real coding; S = stub only.

10.3 Top-down implementation

As mentioned earlier, top-down design is a modular design strategy that specifies creating a systems design in terms of major functions, which are decomposed into more detailed functions. Top-down implementation, which is a general implementation strategy,[2] suggests that high-level modules should be coded and tested before detailed specifications have been completed for low-level modules. High-level modules can be tested by using stubs to simulate the work of lower-level modules. The incremental testing approach is an important aspect of top-down implementation; it serves to distribute integration testing throughout a project and to control the overall testing process.

There are various forms of top-down implementation. For example, top-down purists advocate radical top-down implementation, which involves designing, coding, and testing only one level of a modular hierarchy at a time. Others take the conservative approach, which involves building the entire structural design before coding and testing any modules.

Between these two extremes is a whole spectrum of available strategies, the choice of which should depend upon the characteristics of the problem. In most cases, the conservative approach to implementation has advantages over the radical approach, which if applied literally, often is impractical. For example, if the users wanted to see real output before the entire system has been completed, the radical approach would not work because low-level, detailed modules that get and put real data would have to be implemented before other parts of the system. Or, if developers designed and wanted to evaluate the progress of a project's implementation based on the percentage of completed modules, they could not do so using the radical approach because the structural design, being incomplete, would not indicate the total number of anticipated modules in the system.

Another weakness in radical top-down implementation is that the approach does not provide an early picture of the overall design of a system. Having designed only the top-level modules, the designers really do not know what is coming next. For ·example, if lower-level modules subsequently need information not available from higher-level modules that already have been coded and tested, these higher-level modules will have to be revised and retested. This problem can occur in a conservative top-down implementation environment as well, but in such an environment the structure chart provides design reviewers with an early, clear picture of intermodular communications, down to even the lowest, most detailed modules, before coding and testing begin. As a consequence, the conservative approach minimizes recoding and retesting due to unanticipated problems or inconsistencies in the design.

To show how structure charts can aid in developing an implementation plan and in controlling subsequent implementation, consider the following situation:

1. The users want to know how long implementation will take if two programmers, who each code at the rate of 25 debugged statements per day, are provided to the project.

2. The users want to know how actual implementation time measures up to estimated implementation time.

3. The users want to see the design for the whole system, but they want the system delivered in phases, not all at once. First, they want quick proof that the system is going to work even if no real output is delivered. Next, they want to replace their manual data entry and update procedures with an automated edit and update capability. Then, they want a new automated accounting system and, finally, a forecasting system. They are willing to see any other intermediate versions that the designers and analysts deem advisable.

Assume that structure charts show, in fact, that the system is comprised of four subsystems, each of which does editing, updating, accounting, and forecasting, as shown in Fig. 10.2. Assume further that each subsystem is comprised of forty functionally cohesive modules, and that historical data on other projects have shown that projects with modular designs have modules averaging fifty statements each.

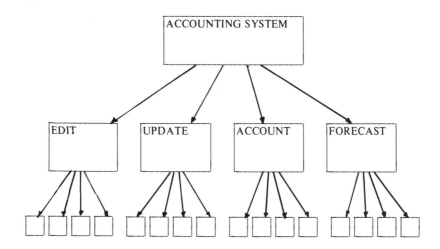

Figure 10.2. A system with four subsystems.

To estimate how long system's coding and testing should take,[3] divide average number of debugged statements per day into average module size. This will give number of days required to code a module.

$$\frac{50\,statements/\,module}{25\,statements/\,day} \;=\; 2\,days/\,module$$

Then multiply number of days per module by number of modules in the system, as shown by the structure chart, to determine number of days required to code and test the system.

$$2\,days/\,module \times 160\,modules \;=\; 320\,days$$

Divide number of days by number of available, full-time programmers to determine implementation time.

$$\frac{320\,days}{2\,persons} = 160\,total/\,days^*$$

To measure actual versus estimated implementation time, consider this: If programmers are assigned to code and test specific modules, the percentage of completed modules at any time can indicate whether the schedule is being met and, if not, to what extent it is overrun.* Again, the structure chart is a useful tool in showing the interrelationships of modules and in developing the order in which modules should be tested to produce the most cost-effective implementation (see Fig. 10.1 and Table 10.3).

In response to the users' third requirement, we again turn to the structure chart. Since it provides a picture of the modular design of the entire system, all parties can visualize how the system can be apportioned to be delivered in phases. For example, users, analysts, and designers might agree that the system will be delivered in six versions, described below:

- Version 1 will prove that major interfaces work within and between the EDIT and UPDATE subsystems.

- Version 2 will show the actual EDIT and UPDATE outputs for only new customer transactions.

- Version 3 will deliver the entire EDIT and UPDATE subsystems.

- Version 4 will prove that all major interfaces of the ACCOUNTING subsystem and of the FORECASTING subsystem work.

---

*The validity of this estimating process is directly related to the functional independence of modules and the absence of high coupling in a program or system. Highly coupled modules and systems with low levels of module cohesion will take longer to implement than the sum of the individual module estimates.

- Version 5 will deliver the actual ACCOUNTING subsystem.

- Version 6 will deliver the actual FORECASTING subsystem.

To meet these objectives, we might decide to code and test the top two levels of the EDIT and UPDATE subsystems to prove that the major interfaces of the system fit together. Then, to implement Version 2, we might determine that some additional modules in the EDIT and UPDATE subsystems would be required. To implement Version 3, we would code and test all of the modules in the EDIT and UPDATE subsystems. (See Fig. 10.3.)

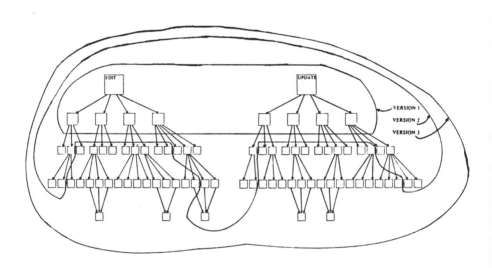

Figure 10.3. An implementation in versions.

This process of using the structure chart to identify the modules required for the first three versions of the implementation can be repeated for subsequent versions. Note that the modules required for Version 1 are only on the top-most levels of the hierarchy, while the modules required for Version 2 do not relate to specific levels in the hierarchy. In actual practice, intermediate versions of a top-down implementation frequently will require the early coding and testing of low-level modules, which get and put the real data that the users are so eager to see.

An implementation plan that calls for a program or system version, which gets and puts real data but does not do much logical processing with the data, typically has an umbrella shape (see Fig. 10.4).

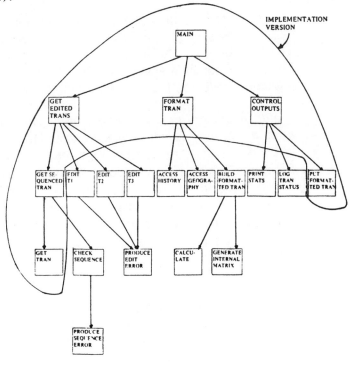

Figure 10.4. Umbrella-shaped implementation plan.

As shown by the above figure, a formatted transaction can be produced for user review and utilized in other parts of the system relatively early in testing. The detailed work of editing, error-handling, calculating, printing, and logging can be deferred by the use of stubs.

From the preceding discussion, we can summarize the advantages of top-down implementation over classical testing approaches:[4]

1. *Unit, integration, and systems testing all are eliminated as separate phases.* In effect, every time a new module (or small group of modules) is added to the system, an integration test is run. The addition of the last module in actuality represents the final system's test. Volume testing and parallel testing, if applicable, follow.

2. *Major interfaces in the system are tested early with a top-down, incremental testing strategy.* Hierarchical modular designs identify high-level modular functions, which can be coded and tested before detailed specifications have been developed for lower-level, detailed functions. As a result, high-level program functions and inter-program system's functions can be tested early, minimizing the likelihood of interface problems necessitating revision of lower-level modules.

3. *Users can see a working demonstration of a skeleton version of the system long before the entire system has been completed,* thereby enhancing users' support for and involvement with the system. Users also tend to have more faith in an EDP department that shows early results, and they can more easily envision how they can use the system. As a result, users more easily can verbalize suggestions for system's revisions. At this early stage, these changes frequently can be

accommodated without revisions to cost and schedule estimates. In addition, users can use early versions of the system to begin training their associates.

4. *Deadline problems are more manageable.* Serious design flaws are exposed early in testing; project monitoring is done small step by small step, and by percentage of project completed; and, as stated above, users can make some productive use of the system before the final version has been completed.

5. *Programmer morale is improved.* Because deadlines are less of a problem and because users have more confidence in the system being developed, programmers are less subject to excessive pressures. Programmer morale also is enhanced because top-down incremental testing provides immediate feedback and satisfaction.

6. *Debugging is easier.* In a bottom-up, phased-testing environment, some modules that have been successfully unit-tested are thrown together for an integration test. If the test fails, the bug could be anywhere in the system, causing considerable difficulty in debugging. With incremental testing in a top-down environment, a module or small group of modules is added to a working skeleton of the system. If the test fails in this case, debugging probably would be reasonably simple, because the bug would exist either in the newly added code or in the interface between the new code and the already tested, working skeleton of the system.

7. *The need for test harnesses is eliminated.* Whereas in a bottom-up environment, testing low-level modules requires driver programs and specialized test data, the actual logic of high-level modules, in a top-down environment, drives lower-level modules. Module stubs simulate the

work of even lower-level modules. Thus, no driver program or specialized test data is created only to be discarded later.

Table 10.4
Machine Time Requirement Comparison

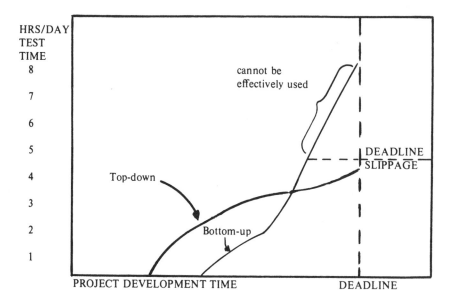

8. *Requirements for machine time are distributed more evenly and are more predictable.* Table 10.4 shows a typical breakdown of machine-time requirements for each of the top-down and bottom-up approaches. In bottom-up implementation, testing begins, perhaps, halfway through a project and grows slowly as modules are compiled and unit-tested. The amount of test-time expands, as the table shows, as interface problems are encountered during subsystem's and

integration testing. Machine-time requirements frequently increase as the project progresses, until the amount of test-time required to meet a deadline surpasses the amount of time available. The reasons for this are summarized as follows:

- Internal priorities. In an environment in which a machine handles both testing and production, the latter normally has the higher priority.

- Type of bugs. As testing progresses, programmers have increasing difficulty finding and fixing bugs, because bugs become more complex. As a result, the mean time between test shots increases because programmers spend more time debugging and recoding than in actual testing.

- Programmer stamina. As systems implementation progresses and the amount of time to recover deadline slippage decreases, the burden of meeting the deadline (or its extension) falls more heavily on programmers. They often are asked to work overtime, but can do so effectively only for short periods of time. But when days stretch into months, the limits of programmer stamina may be exceeded, resulting in loss of critical personnel (if they quit or get sick) or ineffective utilization of time.

In top-down implementation, testing begins sooner than in a bottom-up approach. The amount of daily test-time increases rapidly early in a project because most of the hardware and software must be in place to implement preliminary versions. The amount of test-time then increases gradually as modules are added incrementally to the system. On a New York Times project, for example, the usage of machine time was relatively constant from the ninth month to the twenty-second month when the project was completed. [5,6]

10.4 Packaging and optimization

Packaging is a term used to describe the transformation of logically defined processes into physical units, such as programs, modules, overlays, segments, job steps, and partitions. Systems packaging, which is subject to change, should be defined during the physical design of the system. The objective is to assure that the components of the system are arranged so that available space is not exceeded, execution time is minimized, and the system is easily maintainable. Problems in any one of these areas may require optimization, a repackaging of the system's solution.

A good starting point for a packaging and optimization strategy is to again assume that simple usually is equivalent to efficient. This means we should have manageably small, functionally independent modules that do one, and only one, function. On the other hand, we would not want a system with 20,000 functionally independent modules, each with only five instructions, because it probably would lead to extreme inefficiencies. Neither would we want functionally independent modules that must pass unnecessary switches up and down the modular hierarchy, nor twenty functionally independent modules reporting to one managing module. In all of these cases, we would want to optimize by regrouping functions.

On a systems level, for example, we might have to decide how to group two basic systems functions: editing and updating. One consideration may be operational constraints. If the operating environment cannot support a program that runs for a long time, consumes a lot of memory, and utilizes many input-output devices, then we might want to break the editing and updating functions into two programs. If the only operational constraints were program size and limited I/O device availability, we might break the functions into two job steps within one program. Both of these solutions, however, involve the creation of a type of intermediate file, containing the edited output that also is the update input. If there were no operational constraints, this intermediate file could be eliminated, and we substantially could reduce the number of physical I/Os required by packaging the edit and update functions into one program with one job step.

On either a systems or program level, modular cohesion should be the first concern for packaging and optimization because it leads to low maintenance costs. The advantages of strong modular cohesion should be weighed against other characteristics of the overall modular design structure. For instance, span of control, scope of control/scope of effect, data and control communication, and interface complexity problems frequently can be minimized or overcome by reorganizing parts of the modular structure of the system. Other characteristics of a modular design also can be analyzed to determine the best arrangement of modules. For example:

- Isolate in separate packages functions that are executed only once per program (initialization, housekeeping, termination).

- Isolate functions that are executed infrequently during the program (checkpointing, restarting, special error-handling).

- Group functions with high intercommunication volume (i.e., two functions required every time a record is processed).

- Group functions with a short interval between activation (A record is sequence-checked immediately after it is read, for example.).

- Group functions when one of the functions involves only a few instructions (such as READ and SEQUENCE-CHECK, UPDATE and WRITE).

- Group functions when a module has only one subordinate module.

- Group functions when an executive module manages loop control and a subordinate module controls the body of the loop.

- Avoid excessive creation of intermediate files.

- Sort to provide a natural point of separation.

After the modular design of the system has been completed but before it has been implemented, neither analysts nor designers may be convinced that the packaging of the system meets the time and space objectives of the users. A common mistake is to combine modules prematurely in anticipation of potential bottlenecks. The real problem areas, however, normally turn out to be between or within modules, in such a way as to surprise analysts, designers, and programmers alike. We tend to believe that the system will spend most of its time executing the processes on which we concentrate the most time analyzing, designing, and programming. Because this is a false supposition, we should concentrate, instead, on retaining the functional independence of modules. If optimization is required, programmers can combine simple, separate entities more easily than they can extract functions from multi-functional entities in a complex modular structure.

Until a system has been implemented totally, there may be insufficient data or insufficient physical structure to indicate whether the system's size and speed exceeds its constraints. When determination of response or execution time is critical, we can implement the major interfaces of the system and simulate the work of low-level modules by coding timing loops in module stubs. Although providing only an estimate, this process does indicate whether timing objectives are reasonable.

Nowadays, few modular systems designs necessitate extensive optimization; in most cases, it hardly matters whether a system runs 100 minutes or 105 minutes per week, so long as it is easily maintainable. When optimization is required, most of the processing time usually involves a relatively small amount of the coding. For example, in a program composed of 200 modules, we might expect only 10 percent of the modules to consume 80 percent of the processing time. Knowing this, we can optimize by tuning selected modules of the 10 percent category, rather than by making major revisions to the system's design. Normally, execution time can be significantly reduced by combining modules to eliminate linkage overhead or by rewriting to make critical modules more efficient.

In summary, optimization should be deferred until a system's inefficiencies become apparent or until they threaten the viability of the system's product. Suggested guidelines for optimization are listed as follows:

1. Assure that the system's design partitions, or even over-partitions, the problem into functionally independent modules.

2. If response time or execution time is critical, assure development of an implementation plan that calls for timing simulation in low-level module stubs and simulation of anticipated system's volume.

3. When enough modules have been implemented to provide a sufficient basis for evaluation, determine whether the system meets the time and space objectives of the users.

4. If optimization is required, determine the execution time for each module or cluster of modules.

5. Examine each module or cluster of modules that involves a large percentage of the system's execution time to estimate the largest possible improvement.

6. Estimate the cost required to make the potential improvement.

7. Establish priorities for improvement, taking into consideration the costs of tuning modules, the degree of improvement to be derived, and the expected lifetime of the system.

8. Optimize the highest-ranked modules or group of modules by combining and tuning selected modules, rather than by changing the system's structure.

Review Exercise

1. What is the incremental testing approach? How does it fit in with top-down implementation? How does it differ from the classical testing approach?

2. What are the differences between top-down and bottom-up implementation?

3. What are the advantages of top-down implementation as compared to the bottom-up approach?

4. What are the differences between radical and conservative top-down implementation?

5. What are the advantages and disadvantages of adding new people late in a project?

6. How can a structure chart be useful in estimating the time and costs involved in programming implementation? What other factors should be considered?

7. What are interface bugs and why are these serious?

8. What should be included in a top-down systems implementation plan?

9. Why should non-programming activities and responsibilities be included in an automated systems implementation plan? What are typical problems that might occur if these activities and responsibilities are not defined?

10. What is a modular test plan?

11. What is a versioned implementation plan and what are its advantages and disadvantages?

12. What is meant by umbrella-shaped implementation plan? When might it be employed? What are its advantages over other approaches to implementation?

13. What is the difference between packaging and transforming data flow diagrams into structure charts?

References

1. F.P. Brooks, Jr., *Mythical Man-Month* (Reading, Mass: Addison-Wesley, 1975).

2. E. Yourdon, *Techniques of Program Structure and Design* (Englewood Cliffs, N.J.: Prentice-Hall, 1975).

3. J.D. Aron, "Estimating Resources for Large Programming Systems," *Software Engineering Techniques,* NATO Scientific Affairs Division, Brussels 39, Belgium: April 1970, pp. 68-79.

4. E. Yourdon and L.L. Constantine, *Structured Design: Fundamentals of a Discipline of Computer Program and Systems Design,* 2nd ed. (New York: YOURDON Press, 1978), pp. 340-358.

5. F.T. Baker, "Chief Programmer Team Management of Production Programming," *IBM Systems Journal,* Vol. 11, No. 1, pp. 56-73.

6. _____, "System Quality Through Structured Programming," *AFIPS Proceedings of the 1972 Fall Joint Computer Conference,* Vol. 41, Part 1, 1972.

11 Problem Definition, Objectives, and Estimates

The problem is that the problem keeps changing.

11.1 Problem definition

No matter how systems analyst's functions have evolved or how project personnel are organized, basic problems subvert our systems development efforts time and again. The first challenge is to identify the problem, yet most systems analysts do not have a precise strategy to accomplish this inevitable task. If asked to draw a hierarchical model to show their approach to problem definition, most analysts will respond with a model similar to that in Fig. 11.1, on the following page.

While the model shown in Fig. 11.1 is not wrong, it is too general to assure proper problem definition. Analysts certainly should gather information from users, but analysts traditionally have not identified and spoken to *all* of the *relevant* users. The key question that the systems analyst should ask is, Which users have *all* the information that should be gathered? By gathering information from all relevant users, the systems analyst can better assure development of an acceptable solution because he will have defined the whole problem, not just 95 percent of it.

Analysts should begin the task of problem definition by returning to the source of the problem evaluation request. All users in departments who prepare the inputs to and receive the outputs from the current system also should be interviewed. These users, however, frequently do not comprise the entire potential user community. For example, standards, operations, quality control, and future planning departments also might be concerned parties regarding the development of new systems.

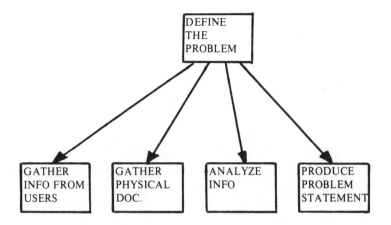

Figure 11.1. Typical approach to problem definition.

One way to help to identify the total user community is to determine the organizational relationships and responsibilities of all departments in the user environment. Organizational charts are useful to define departmental relationships but they frequently do not clarify responsibilities. For example, an organizational chart may indicate that four of the thirty departments in the organization are Systems Review, Technical Review, Systems Acceptance, and Quality Assurance. From their names alone, these departments' functions are unclear, as well as their responsibilities and relationship to the problem or its potential solutions. Systems analysts should identify all of these things and, in the process, pinpoint the key contact for each department.

When the key contact within a user department says to an analyst, "If you have any questions, speak to me but do not bother my people," the analyst has a problem. If the analyst accepts this limitation, incomplete problem definition is almost a certainty. For this reason, we should convey to the user that within any department there often are different levels of user, each with unique perspectives, responsibilities, and insights. For example, strategic management only may be concerned and most familiar with the general direction and profitability of the organization. Middle management may be interested in meeting departmental budgets within existing manpower constraints.

Lower-level workers, however, probably are most familiar with the details of the current system's environment, including its weaknesses and strengths. Each level of user should be interviewed, for each can provide worthwhile feedback: Disregarding any potential source of information will limit the chances for successful problem definition.

Of course, asking the right questions and recognizing when those questions are not being answered properly are crucial to success. Too often, systems analysts attend problem definition meetings, only to realize days later that the problem still has not been defined. It is important, then, for the systems analyst to know what information he needs. He should be aware that a well-stated definition should distinguish the problem from its causes, effects, related considerations, and potential solutions.[1] Furthermore, the definition should answer the who, what, when, where, why, and scope of the problem. The systems analyst must be able to obtain and process information from all user levels; he should not rely on just one user to be a good judge of the proper solution to develop.[2]

Too many analysts start with the solution and work backward to a definition of the problem. An example of this might be the organization that initiated a pilot project utilizing structured techniques: After a few months' progress, a consultant was brought in. He asked, "At what stage is the project?" The answer was that the problem definition phase had not been completed, but project management already had decided that the system would contain three programs.

In another case, a directive had come from the top of an organization to optimize testing. The analyst assigned to the task went from department to department trying to identify all of the complaints associated with the testing environment. The more people he spoke to, the more complaints he got, most of which were petty, but the majority of users readily admitted that the current testing environment was far superior to the previous one, and advised against optimization. After months of working backward to a definition of the problem, the analyst discovered that there was no problem at all.

Normally, the solution should fit the problem, and we should work from the problem to the solution. In some instances, however, systems analysts should be prepared to fit the problem to the solution. For example, sometimes users know what they want or need, such as an output report. (Sometimes the directive is a requirement, not just an objective, such as when the government says, "You will.") In this case, the systems analyst's task is to work backward from the content and format of the report to the sources of the data.

After speaking to all users in the first phase of problem definition, systems analysts should gather all written documentation related to the problem. Such documentation frequently includes a definition of manual procedures; details of an automated system's solution such as program, module, file, and record descriptions; and performance reports that indicate volumes, error rates, and overall operational efficiency of the current system. Analysts also should identify any future plans that may exist in the form of reports and memoranda from research and planning departments. Unfortunately, studying the physical systems documentation alone usually does not provide analysts with a clear understanding of the current environment, because it very often is incomplete and out-of-date. At other times, so much technical detail in the documentation can obscure the basic functions and limitations of the system.

In order to understand the current system's operations, the analyst should develop a data flow diagram to derive the logical flow of data and logical processes required to transform inputs to outputs. While users who are "results-oriented" may balk at the time taken to study the current system, the effort is cost-justifiable in the long run.

Regardless of specific paper, card, tape, or disk inputs and outputs, the analyst should develop a basic functional systems flow to identify missing, duplicate, extraneous, and relevant functions of the current system. By doing so, the analyst will be able to compare the functions of the current and proposed systems and to identify and quantify the additional functions that will be required.

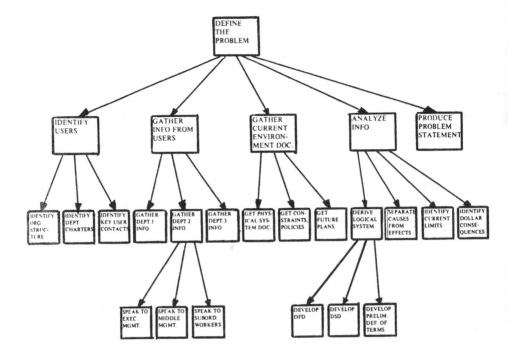

Figure 11.2. Disciplined approach to problem definition.

From the above discussion of typical pitfalls in the data gathering and evaluation phases, we can develop a more complete and precise model for problem definition, as shown in Fig. 11.2. In addition to the functions established in Fig. 11.1, the systems analyst should

- Get an organization chart.

- Identify all departmental responsibilities and functions.

- Identify the primary user contact within each department.

- Talk with various levels of users within each department.

- Access performance reports and physical documentation and identify future plans.

- Derive logical systems documentation.

- Separate causes from effects.

- Identify current systems' limitations.

- Identify and quantify current systems' costs.

11.2 Defining objectives

Having defined the problem, the systems analyst next must identify the objectives of a new system. Although large systems usually serve many users with different perspectives and objectives, let's simplify the problem and assume that a system is to be developed for only one user. Sooner or later, the analyst will ask that user to define his objectives. The user might say

"I don't care how you do it, but I need valid information faster. Also, if we're going to increase profits, that information had better be more accurate than the garbage I'm getting now. That's the whole story."

But is that the whole story? Too often from such a conversation, the analyst thinks he has obtained a clear picture of the user's objectives — to improve accuracy, to increase profit, and to provide faster retrieval of information. He goes back to the drawing board and proposes a system to meet these objectives. The user, not knowing much about systems alternatives and thinking that the analyst understands his problems and objectives, accepts the systems proposal.

Systems design and implementation proceed. Three months later, the user calls the analyst and asks, "When am I going to get this faster information retrieval?" The analyst answers, "We've tested personnel transaction records and credit transactions, and next week we'll start testing the statistics part of the system. We think you are really going to like it." After a long silence, the user responds, "You had better get over here immediately. I don't know what you're doing, but it's certainly not what I want."

From the scanty detail in the above story, it is not clear what the user wants that the analyst is not providing. It is clear, however, that there has been a failure in communications from the start, with premature acceptance of the user's statement of objectives. Although the user may have been at fault in approving a systems proposal that he did not fully understand, the primary fault is with the analyst in not probing the user's statement sufficiently to determine his specific objectives.

A well-defined objective must be measurable. There should be no misunderstanding between parties as to the objectives or whether they have been met. Three considerations are the time and cost to produce a result and a clear statement of the result required. The analyst in the preceding story should have asked the following questions:

By how much do you want accuracy improved, based on what figures?

What specific information do you need faster, and how much faster do you need it?

How much are you willing to spend to achieve these objectives?

How quickly do you need the job done?

Had the analyst asked these questions, he might have gotten an entirely different picture of the problem and the user's objectives. Perhaps, he would have determined that the user actually wanted the following:

Within three months and at a cost of no more than $20,000, improve accuracy on inventory transactions by no less than 20 percent using the data from the June 1976 report as a base.

Instead, the analyst in our example was plagued with the same type of vagueness that too often is manifested in problem definitions, statements of objectives, and various levels of specifications: In the user's statement of objectives presented at the beginning of this section, note the unclear adjectives (What is *valid* information?), unclear adverbs (What is meant by need information *faster?),* and imprecise verbs (What do you mean by *increase* profits?).

Here is another variation on the same theme. The user says, "Reduce error rates." The analyst returns with a system that has reduced the error rate from four percent to one percent, to which the user says, "That's not what I wanted." The analyst asks, "What do you mean? You told me to reduce error rates, and they have been reduced substantially." "Not good enough," the user says. "My threshold for acceptance was an error rate of four-tenths of a percent." The analyst then says, "Why didn't you tell me?" and the user responds, "Why didn't you ask?"

If this were an unusual, isolated story, perhaps we might be inclined to place equal responsibility for imprecise definition on both parties. Instead, this communication gap can be expected in many aspects of systems development efforts. The users involved may be new to data processing, but systems analysts, for the most part, are war-scarred veterans who have been through the battle of defining objectives before. They should know better, and it is their job to ask the probing questions to get the precise answers they need.

11.3 Estimates of costs and benefits

Systems traditionally have been delivered late and over-budget. As systems development efforts increase in size and complexity, the overrun in budget can be so large as to cripple an organization. Although large systems development is like trying to build a house underground, management is under severe pressure to quantify the total cost of the development effort. That pressure sometimes is so severe that management immediately translates an estimate from an analyst to a fixed figure. Too often, an analyst, when pressed for an estimate, may respond with the following:

"If we freeze the specifications today (usually unrealistic), and if we don't run into substantial problems (perhaps even more unrealistic), then we can produce the system for $1,000,000, plus or minus a fudge factor of 15 percent."

Unfortunately, by the time top management in the organization hears the story, the estimate for the system is simply $1,000,000.

Analysts cannot be held responsible for the way in which their estimates are translated by others. They should, however, be aware of these management pressures and, as a result, make absolutely clear not only what they know, but also what they do not and cannot know. For example, while management may not be pleased with "I can't define it better because . . . ," they will at least have an early awareness of the possibilities and degree of potential budget overrun if the reasons are genuine and well-stated. Some of the problems inherent in making estimates for a systems development project follow:

1. *Methodologies may call prematurely for estimates of total systems development costs.* Some methodologies, for example, start with an initial data gathering phase to define the problem and the feasibility of continuing the project. Of course, this phase is limited in time and cost factors, because management does not want to spend, say, $50,000 only to find that the project is not feasible. Yet, the methodology may require the feasibility statement to define an estimate of total systems development costs and benefits. One problem with this approach is that the costs and benefits of different solutions can range from a quick-and-dirty solution costing $20,000 and meeting 80 percent of the users' objectives, to a $2,000,000 solution that meets 99 percent of the objectives. Even an estimate that is understood to cover a wide range of solutions (plus or minus 300 percent, for example) may not cover the range in potential costs involved in these solutions.

    Another problem is the limited time available for the analyst to make the first estimate of total systems development costs. The question is, How can the analyst determine this estimate before he has spent considerable time defining objectives, priorities, and potential solutions involving tentative physical design? The answer, of course, is, He can't. Moreover, his responsibility is to say he can't, and to explain why.

2. *Many financial factors are beyond the control of the analyst who is making the estimate.* For example, if the system will utilize a software package or hardware developed by an outside firm, obviously, the price of that software or hardware may be subject to change. Another example is inflation. The analyst may project programmer costs at the rate of $20 per man/hour, as dictated by his organization. A year later, when the system is beginning to consume large chunks of manpower resources during the implementation phase of the project, inflation may have driven the man/hour cost to $25. To anticipate such problems, the analyst should, at least, document the assumptions used to develop his estimates, and for large projects of long duration, include some inflationary factor in estimates.

3. *The problem is that the problem keeps changing.* If the systems analyst has not correctly identified the user community, for example, sooner or later he will find out that the problem that he has been trying to solve is not really the whole problem. Thus, any estimates already developed will have to be redone. In another example, as the system is being developed, the users recognize its potential and want enhancements to the system — all, of course, to be done as soon as possible. To incorporate these enhancements, the system's design and estimates will have to be modified accordingly. Obviously, the analyst cannot prevent the users from changing specifications, but he must make clear to the users that as the problem changes, so will the estimates.

4. *The analyst does not realize the complexity of the job.* Does the job involve application or systems programming? Is the job routine or innovative? Statistics show that programmer productivity in terms of the number of lines of code produced per day varies greatly according to the type of programming required. In one study, the difference in pro-

grammer productivity between easy and difficult coding showed a four to one difference in the number of instructions coded per man/day. [3]

Another question for the analyst to consider is, What will be delivered? Will it be a program, a system, a product, or some combination of the above? As Brooks so vividly points out, [4] generalizing a program so that it can be used in many operating environments and can be fixed or modified by numerous people, increases the development costs by a factor of three. Similarly, designing a program so that it can interact with other programs in a system requires, perhaps, three times the development costs of a stand-alone program. Finally, a programming system's product is more complex than either a program product or a programming system, and its development may cost nine times that of a stand-alone program. See Fig. 11.3.

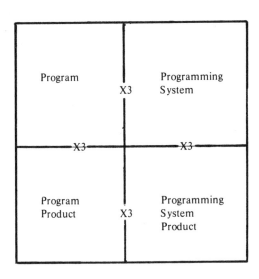

Figure 11.3. Relative developmental costs by project type.

5. *Analysts (who should be pessimists by this time) tend to be eternal optimists.* One problem is that they continually underestimate the effect of communication and coordination problems on the development of large systems. Another prob-

lem is that most analysts who develop estimates have considerable experience in the field and unconsciously project themselves into the implementation phase of a project. In most cases, however, people having little experience with the application, equipment, programming language, standards, policies, and terminology of the organization may actually do the implementation. In determining estimates, the analyst should identify the actual experience of the people who will be doing the job. Are they senior- or junior-level workers or trainees? Are they of average or extreme competence? Will they be committed to the job on a part-time or full-time basis? These questions must be answered when the analyst is developing estimates so that the estimates will prove reliable.

6.   *Analysts often do not have access to historical data that can be used as guidelines for estimating costs.* While each systems development effort has unique characteristics, frequently other organizations or project teams have attempted similar efforts. Although the analyst would find it useful to examine the costs and problems involved in such efforts, too often the data are not readily available or the analyst does not dig deeply enough to get at this information. Although we surely cannot expect one insurance company, for example, to help a competitor, there is no reason why the State of California should not ask the State of Texas about its experiences with a particular application. Whenever possible, the analyst should gather data from similar projects to determine the costs involved in training, programming, testing, software packages, and hardware.

7.   *Typically, estimates have reflected only developmental costs.* Even if a systems development effort meets its developmental cost estimates, the users may not be happy with the system, because they may spend more money maintaining and enhancing the system than in developing it. In developing estimates, systems analysts should include estimates of costs for the system's lifetime.

An estimate is not complete unless it identifies anticipated deviations. Any analyst who says in a preliminary meeting that a system will cost one million dollars, no more and no less, probably should be reassigned immediately to a less critical path. A well-stated estimate should contain a range of values, the variables considered in developing the estimate, and a confidence factor associated with the extremes of the estimate.

	1	2	3-4	5+
1. Size of team	☐	☐	☐	☐
	1	2	3	4+
2. Number of users	☐	☐	☐	☐
	on team	average	little	remote
3. Availability of users	☐	☐	☐	☐
	well-defined	average	unclear	none
4. Understanding of users	☐	☐	☐	☐
	simple	average	complex	unknown
5. Complexity of application	☐	☐	☐	☐
	senior	average	junior	trainee
6. Experience of team personnel	☐	☐	☐	☐
	0-10%	11-25%	26-50%	over 50%
7. Time on other work	☐	☐	☐	☐
	well-defined	average	unclear	none
8. Availability of historic data	☐	☐	☐	☐

Figure 11.4. Questionnaire for estimating.

For example, assume that one analyst has developed a preliminary $500,000 estimate for the development of a system without an accompanying description of the detailed characteristics of the project. Because the estimate is misleading (it is too precise) and incomplete (no rationale is given), the first analyst asks a second analyst to develop a well-stated estimate. By utilizing a questionnaire such as that shown in Fig. 11.4 and by attaching numeric values to answers as shown in Fig. 11.5, the second analyst might determine that the original estimate should be considerably higher.[5,6] If, for example, he determined the project had the characteristics as shown in Table 11.1, the refined estimate would be $550,000 to $650,000, depending on the experience of the people who will be put on the job. By presenting an estimate in this way, the analyst gives management sufficient information to confront pressures from their superiors. Management could say, for example, that if senior personnel were assigned and if better user availability could be established, then the original $500,000 estimate could be met.

Size of team		Complexity of application	
1	= −.2	simple	= −.2
2	= 0	average	= 0
3–4	= +.1	complex	= +.2
5+	= +.3	unknown	= +.4
Number of users		**Experience of team**	
1	= −.1	senior	= −.2
2	= 0	average	= 0
3	= +.2	junior	= +.2
4+	= +.4	trainee	= +.4
Availability of users		**Time on other work**	
on team	= −.2	0–10%	= −.2
average	= 0	11–25%	= 0
little	= +.1	26–50%	= −.2
remote	= +.3	above 50%	= −.3
Understanding of users		**Availability of historic data**	
well-defined	= −.1	well-defined	= −.1
average	= 0	average	= 0
unclear	= +.2	unclear	= +.1
none	= +.4	none	= +.2

Figure 11.5. Numeric values relating to questionnaire answers.

Table 11.1
Project Profile for Estimating

Characteristic	Detail	Value
Size of team	3–4	+.1
Number of users	3	+.2
Availability of users	average	.0
Understanding of users	well-defined	−.1
Complexity of application	complex	+.2
Experience of team	senior or average	−.2 or .0
Time on other work	0–10%	−.2
Availability of historic data	unclear	+.1
Original estimate	$500,000	1.0
Refinement factor		1.1 or 1.3
Refined estimate	$550,000 to $650,000	

Keep in mind that dollar estimates refer to both costs and benefits — that is, better service, increased profits, decreased costs, among others — and that costs are justified in terms of benefits.[7] If the analyst has done his job properly, there usually are many possible solutions to a particular problem, each solving the problem to some degree at some cost. The analyst should present an estimate of both costs and benefits for each proposed solution, noting the degree to which the problem will be solved. For example, assume that users of a proposed system have established a requirement to access certain information. While the users would like this information as soon as possible, they may not want to pay the costs involved in immediate retrieval of the information. The analyst should produce a cost/benefit analysis for alternative solutions, identifying the incremental value associated with each. Because most users are not interested in many details, a simple graphic, such as that shown in Table 11.2 on the following page, can illustrate the important considerations. As you can see, the table summarizes some of the critical monetary points of each alternative, and more importantly, enables users to compare easily the costs and benefits of each alternative.

Table 11.2
Cost-Benefit Analysis

ALTERNATIVES	RES-PONSE TIME	% OBJEC-TIVES MET	DELIVERY TIME	BENEFIT/ YR ESTI-MATE	DEVELOP-MENT COST ESTIMATE	VALUE/YR ESTIMATE (5 YEARS)*
1. 100% manual	48 hr.	60%	1 mo.	30,000±20%	15,000±20%	84-$156,000
2. 50% manual, 50% batch	24 hr.	70%	4 mo.	50,000±20%	50,000±20%	80-$220,000
3. 100% batch	24 hr.	80%	6 mo.	100,000±20%	80,000±20%	208-$472,000
4. 50% batch, 50% real-time	60 sec.	90%	1 yr.	150,000±20%	170,000±20%	192-$628,000
5. on-line data base	10 sec.	99%	2 yr.	200,000±20%	300,000±20%	80-$720,000

*Includes provision for maintenance and modification costs over five years assumed to be equal to original developmental costs.

Review Exercise

1. How can an organization chart help to define a problem?

2. How can knowing a department's functions and responsibilities help to define a problem?

3. What is meant by the phrase, "talk with all levels of the user community"?

4. Under what circumstances would you work backward from the solution to the problem?

5. How does the concept of logical analysis (i.e., what before how) relate to problem definition?

6. Why might an analyst have to derive the current logical system before defining the problem?

7. List typical pitfalls in developing problem definitions.

8. How would you identify a well-stated objective?

9. Why are detailed characteristics of a project important in the development of well-stated estimates?

10. Give examples of how historical data can help in the estimating process.

11. Why should maintenance and enhancement costs be included in the cost estimates for a system?

References

1. C.H. Kepner and B.B. Tregoe, *The Rational Manager* (New York: McGraw-Hill, 1965).

2. S. Lieberman, "Let the Buyer Be Heard," *Industrial Design,* November-December 1976.

3. J.D. Aron, "Estimating Resources for Large Programming Systems," *Software Engineering Techniques,* NATO Scientific Affairs Division, Brussels 39, Belgium: April 1970, pp. 68-79.

4. F.P. Brooks, Jr., *Mythical Man-Month* (Reading, Mass.: Addison-Wesley, 1975).

5. J. Toellner, "Project Estimating," *Journal of Systems Management,* May 1977.

6. L. Fried, "Estimating the Cost of System Implementation," *System Analysis Techniques,* J.D. Couger and R.W. Knapp, eds. (New York: John Wiley & Sons, 1974), pp. 498-509.

7. J. Emery, "Cost/Benefit Analysis of Information Systems," *System Analysis Techniques,* J.D. Couger and R.W. Knapp, eds. (New York: John Wiley & Sons, 1974), pp. 395-425.

12 Miscellaneous Problems

In most cases it is not the change itself that people find so distasteful but rather how that change has been imposed.

12.1 The problem with having no methodology

Some organizations have not adopted a specific methodology for the development of data processing systems. They either feel that the generalized standards already developed impose a sufficient level of control of activities, or that project development guidelines should be left to the discretion of each project manager. They may also feel that the variety of projects being developed does not lend itself to standardized methodology. The documents to produce, the reviews to conduct, the phases to perform, and especially the amount of detail in all of these efforts should vary, to some extent, according to the magnitude, complexity, and schedule of the specific job. However, without a methodology that specifies the order and detailed steps of these activities, projects can lack structure and control.

An analyst working without a formal methodology will be under considerable pressure to assure that the objectives and priorities of the users are met. He will have to oversee design and implementation efforts to assure that neither strays from the goals of the users or of the business. He will have to monitor the project's progress to assure that schedules and budgets are followed. He will have to assure that documentation is complete and not merely a superficial afterthought. And, finally, he must assure that the users of the system have been properly trained and, therefore, are prepared for the system when it is finished.

12.2 Limitations of current methodologies

With the realization that even the best systems analysts and project managers cannot control large systems development efforts without a formal methodology, many organizations either have developed their own internally or have bought packages developed externally. Both approaches give structure to projects and improve senior management's feeling of control, largely because they impose checkpoints and progress-reporting requirements. However, current methodologies still are limited in important ways, as discussed below:

1. *Most methodologies, while attempting to reduce communication problems, result in overdocumentation.* Building a complex system without any documentation obviously is undesirable, but many modern systems developers have fallen into the trap of specifying that everything produced or decided upon should be recorded in great detail. The assumption is that if everything to be communicated is written in a document and circulated to everybody who needs to know about it, then the document will be read and its contents communicated to the recipients. This is a fallacy.

    A number of methodologies specify in detail the various documents to be produced during each stage of a project, and their tables of contents. Unfortunately, most methodologies do not provide guidelines to assure that these documents are of high quality or that they can be communicated easily. Analysts must see to it that the reviewer can understand the document. In addition, the reviewer must be given enough relevant facts for a decision — without extraneous verbiage to obscure the problem.

    Because the real business world is complex, comprehensive documents about any aspect of that world tend to be thick, sometimes hundreds of pages long. Harassed businessmen too often must review and sign off on such documents without adequate time or technical knowledge to assess their contents. Such documents frequently are the basis of an implicit contract between businessmen and the data processing department. One problem is that the contract may not be legal or enforceable, because the businessmen and data processing employees may work for the same organization. A year later, when a businessman says, "I don't care what I signed; this system doesn't do what I need to have done," the system probably will have to be changed and thousands of dollars will have been wasted.

As analysts, we should be alert to ways by which such misunderstandings can be prevented. One solution is to be found through the graphic tools of structured analysis, which will increase the vividness of a document and decrease the numbers of words that have to be read. As analysts, we also should be aware of potential morale problems inherent in overdocumentation. Whenever there is a redundancy of effort in developing or reviewing documents, we can expect resistance and frustation to result on the part of both developers and reviewers.

2. *The project phases defined by most methodologies are not sufficiently iterative.* Most methodologies assume that once a document is produced and approved that that is the end of it. For example, once a design has been produced and approved, that remains the final design. As a result of this way of thinking, most formal methodologies prescribe a set of nonoverlapping project phases, requiring, for example, a feasibility study of a system to be produced and approved before general design can start; detailed design completed before implementation can start, and so on.

However, the true sequence of events in a typically complex project is more like a spiral than a straight line, as diagramed in Fig. 12.1. In order to determine the cost/benefit estimates for possible system's solutions, some tentative physical design is required. As we investigate the possibilities of design, we may determine that the feasibility study contained some false conclusions. In general, as we start implementing a system, we may find out that there are some details missing from the system's design. As we go deeper into a system's problem, we are likely to gain insights that may require revision of previously accepted documents. Management and technicians should recognize that to go back to a problem to rethink and revise is both natural and desirable.

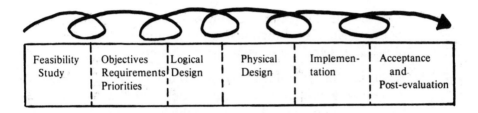

Feasibility Study	Objectives Requirements Priorities	Logical Design	Physical Design	Implemen- tation	Acceptance and Post-evaluation

Figure 12.1. The iterative approach.

This spiral nature of our work may appear very untidy from management's point of view. To maintain control, management would like a firm statement of what has been completed and what needs to be done. Sometimes, both management and the procedures outlined in formal methodologies demand adherence to the straight-line approach to the phases of a project. Consequently, the person actually responsible for doing the work may pretend he is proceeding through a series of independent phases when, in fact, his progress more closely resembles the back-and-forth spiral approach. We should recognize that some people, if allowed to spiral, will spiral indefinitely, never committing themselves to any approach. Nevertheless, the spiral nature of systems development activity should be incorporated into our systems development methodologies.

3. *Current methodologies fail to make a clear distinction between logical systems and physical systems.* Most methodologies imply that analysts and designers can proceed from an understanding of the current physical system to the proposed physical system without defining the logical requirements of both the current and proposed systems. For the systems analyst and the designer, definition of logical requirements provides a desirable intermediate step in the

analysis and design processes. For the user, such definition provides information about data and processes — in non-technical language and at a logical level.

A logical record is a piece or set of information that we want to use in our system. The physical record establishes how that logical record actually will look. The concept of a logical design or functional specification, as opposed to a physical design, is an extension of this concept.

Assume, for example, that we want to specify a logical design of a proposed system by defining the inputs, outputs, objectives, and the way in which the proposed system will transform the inputs into outputs without committing ourselves to specific manual, batch, or on-line processes. The statement *Customer account inquiries consist of customer account identification, authorization code of the inquirer, and transaction code specifying inquiry type* is a completely logical statement. On the other hand, the statement *Customer account inquiry cards have a zoned decimal customer account number punched in columns 2 through 10, a zoned decimal authorization code punched in columns 13 through 16, and an alphabetic transaction code (ranging A through H) in column 20* is a totally physical statement.

If you are in doubt as to whether a statement is logical or physical, envision whether the statement could be physically implemented in more than one way. If the answer is yes, the statement is logical. If, on the other hand, there is no other way in which that statement can be physically realized (as in the case of the statement about customer account inquiry cards with specific locations and formats of fields), the statement is physical.

12.3 The need to distinguish priorities

Current methodologies fail to identify how to distinguish priorities. As pointed out earlier, all pertinent points of view must be solicited in order to design the best system. Unfortunately, the more people we speak to, the more stories we get; and we can't please all of the people all of the time. The analyst's responsibility, therefore, is to get a consensus on the objectives and priorities acceptable to the greatest number of relevant decision-makers.

Because most systems serve many users, the analyst must go beyond merely asking users what they want. The analyst also must determine how much the users want it and what use the system will serve. The analyst must develop effective probing techniques to identify the between-the-lines meaning of statements about proposed systems objectives. Perhaps most important, he must supply the imagination that the users may not have, due to their inexperience with the possibilities of data processing technology.

For example, a user may say, "I want to improve accuracy, get faster response time, and be able to handle more volume." The analyst should refine these objectives not only by quantification, but also by establishing priorities within objectives. When the user says, "This system will help me," the analyst should ask, "How?" When the user says, "I'll get a better picture of where we are," the analyst should ask, "So what?" When the user has trouble understanding what the system could do for him, the analyst should tickle the user's imagination by asking questions like, "What are the major factors that currently stand in your way?" or "How would you do your business differently if you had more or faster information?" or "What kind of information would you want if you could get it?" Keep in mind that people are creatures of habit, and that they may be unaware of the dramatic advancements in potential technological possibilities.

It is the analyst's job to probe, inform, and educate. These tasks, especially probing, can alienate users, but if the job is done diplomatically, the users will be able to appreciate the analyst's problems and his need for specific statements. Constant probing enables the analyst to transform technical features of a system into measurable benefits. Recovering avoidable costs, capturing lost revenue, and improving service to customers all are benefits to be identified that ultimately will justify the technical outlays.

12.4 Controlling detail

Systems development involves accumulation of thousands of details, which should be recorded permanently in the documentation of the system. Analysts traditionally have used the English language in narratives and procedural flowcharts to communicate and document ideas and details, but both English narratives and flowcharts tend to result in communication problems. English narrative is notoriously ambiguous and tends to be verbose. User managers and others who must review documents typically do not have the time to read thick, ambiguous, overly technical documents. Procedural flowcharts usually are either so general that they virtually are useless, or so detailed that they are barely a level of detail above the actual code. Moreover, users do not relate easily to symbols, loops, sequences, and switch decisions depicted in flowcharts.

Recognizing these weaknesses, the analyst should use more reviewer-oriented communication tools. It is almost useless to present a physical system's solution to a nontechnical reviewer, who cannot relate to "computerese." What the user can relate to are his logical requirements. So as a first step toward better communication and controlling detail, the analyst should present a logical design for users' review before any development of a physical system's representation. Second, the analyst should use graphics to summarize on a few pages what would be verbalized in twenty. Third, all terms in the logical design should be defined precisely. Fourth, the document should be segmented so that pertinent sections stand out for specific reviewers.

While some guidelines for a document's contents can be established, perhaps a more important concern is *how* the document is presented. Typically, we have distributed documents to be read, but reviewers can get frustrated when they can't get an immediate clarification on something they don't understand. The analyst should realize the importance of face-to face, sales-type presentations. By means of flipcharts or transparencies, the analyst effectively can present his findings. He can be assured not only that the material has been reviewed, but also that reviewers understand the material, because they can get immediate clarification of points they don't understand. Most important, both the reviewers and the analyst can develop a dialogue from which new ideas can be generated. In this way, important communication channels are opened to the fullest extent.

12.5 Resistance to change

One of the inevitabilities in a systems development environment is that things are changing constantly — users think of new functional objectives of a system after the specifications supposedly have been frozen; new equipment replaces outdated hardware; technicians learn one operating system only to be informed that there is to be a massive conversion to another; and standards and methodology groups constantly change their guidelines to keep pace with the dynamic environment.

Associated with these changes is a persistent and baffling problem: resistance to change, which takes many forms and results in crippling productivity. For example, key people may refuse to be a part of a change and either may quit or take positions outside of their areas of competence. Internal quarrels, hostility, and political maneuvering may result in a breakdown in the collaboration required for effective systems development efforts. And, everything just may seem to slow down because people are not happy about their working environments.

Certainly, people can be expected to resist change. Systems analysts and project management, however, would do well to analyze the cause of the problem, rather than merely treat its symptoms. Perhaps the most important factor in this resistance is people feeling that they are not treated with consideration. If workers potentially are to be affected by a change, they naturally feel they should have something to say about the nature of that change.[1] In most cases, it is not the change itself that people find so distasteful, but rather how that change has been imposed. What people resist "is usually not technical change but social change — the change in their human relationships that generally accompanies technical change."*

---

*Paul R. Lawrence, "How to Deal with Resistance to Change," *Harvard Business Review,* January-February 1969, p. 4.

Get the user involved; get the people to participate; make the effort a collaboration — these all are typical conclusions of an analysis of the resistance-to-change problem. As solutions, however, they may raise as many problems as they solve. Some resistance that is an expression of dissatisfaction perhaps should be viewed as an opportunity for improvement, rather than merely as something to be eliminated. Ultimately, however, the basic problem does not concern the *value* of resistance. Instead, the question is, *How* does one get people involved to participate and collaborate actively on a project?

Organizations can set up strategies and procedures to help assure open lines of communication and active participation, but no artificially created peer review, or meeting of all concerned, can assure effective participation. It involves much more. Workers need a sense of belonging, an attitude of togetherness, and a feeling of mutual contribution. They require an environment in which management and workers respect each other and continually work to cultivate the relationship.

12.6 The ruts we have created

If we tend to resist change, perhaps it is because we feel so comfortable in the ruts we have created. A particularly serious rut concerns our approach to problem-solving. For example, an executive of a major financial institution recently complained to me that people keep trying to solve problems in the same unimaginative way, even when past efforts have proven less than totally successful. His complaint reminded me of a situation that has particular relevance to problem-solving in general, and to systems analysis in particular.

Consider the situation of trying to control insects with pesticides, an approach that certainly causes many problems, perhaps more than it solves. For example, cotton farmers continue to try to eliminate boll weevils by spraying insecticides, which work reasonably well for a time. Gradually, however, the insects have built up a resistance to previously poisonous insecticides, thereby creating a need for new, more effective chemicals.

While this cycle continues to repeat, insect damage to cotton crops is not decreasing, and the environment is being continually contaminated by the chemicals.

How refreshing it was to hear that a new approach to the problem was being pursued in California. Apparently, a study on the habits and preferences of the boll weevil resulted in the fascinating conclusion that the insect did not particularly like cotton. Strangely enough, the boll weevil seemed to prefer alfalfa. So, alfalfa strips were planted between cotton fields to tempt the boll weevil to eat the alfalfa and to leave the cotton alone. I do not know the results (or the source) of this experiment, but what I do find most impressive is the freshness of the approach.

Keeping the preceding story in mind, try to solve the following logic problem:

Al, Bob, Charlie, Don, and Ed came upon some money while hiking in the woods. Because the money had been scattered by the wind, each collected as much money as he could find. When all of the money had been gathered, it became apparent that Al had collected much more than any of the others. Under pressure, Al agreed that he would give up some of his money by doubling the amount held by each of the others. After doing so, Bob had much more money than anyone else. Al commented, "What's fair for me should be fair for you, Bob." So Bob agreed to do what Al had done, resulting in Charlie now having the most money. So, Charlie doubled the amount of money held by each of the others. Since Don now had the most, he doubled the amounts of money. By now, you might guess the result: Ed had the most money, by far, and he agreed to double the amounts of money held by each of the others. What you don't know is that after Ed doubled the money, all five of them had the same amount of money. The question is, What was the original ratio of money held by each of the five before Al doubled their money?

Whenever I present this problem to seminar attendees and associates, they usually cannot solve the problem unless they devote considerable time to determining the correct answer algebraically. Sometimes, people say that this problem reminds them of all the algebra they had forgotten. Others say they lose interest once they realize that the solution involves tedious work to develop, reduce, and substitute some set of algebraic equations.

Yet, some people solve the problem in only a few minutes without using any sophisticated algebra. All the problem requires is a simple, efficient approach, working backward rather than forward. We can assume that in the end, each person has one-fifth of the total amount. Using these fractions as a starting point, we easily can determine the original ratio, as shown in Table 12.1. Or, by assuming that they came upon some arbitrary amount of money, we can solve the problem as shown in Table 12.2 (note that a power of 2 is easy to work with).

Table 12.1
The Money Problem — Solution 1

A	B	C	D	E	State
$\frac{1}{5}$	$\frac{1}{5}$	$\frac{1}{5}$	$\frac{1}{5}$	$\frac{1}{5}$	at end
$\frac{1}{10}$	$\frac{1}{10}$	$\frac{1}{10}$	$\frac{1}{10}$	$\frac{6}{10}$	before E doubles
$\frac{1}{20}$	$\frac{1}{20}$	$\frac{1}{20}$	$\frac{11}{20}$	$\frac{6}{20}$	before D doubles
$\frac{1}{40}$	$\frac{1}{40}$	$\frac{21}{40}$	$\frac{11}{40}$	$\frac{6}{40}$	before C doubles
$\frac{1}{80}$	$\frac{41}{80}$	$\frac{21}{80}$	$\frac{11}{80}$	$\frac{6}{80}$	before B doubles
$\frac{81}{160}$	$\frac{41}{160}$	$\frac{21}{160}$	$\frac{11}{160}$	$\frac{6}{160}$	before A doubles

Table 12.2
The Money Problem — Solution 2

A	B	C	D	E	Total	State
32	32	32	32	32	160	at end
16	16	16	16	96	160	before E doubles
8	8	8	88	48	160	before D doubles
4	4	84	44	24	160	before C doubles
2	82	42	22	12	160	before B doubles
81	41	21	11	6	160	before A doubles

My point in presenting the money problem is that we should be aware of our natural tendency to get into ruts when confronted with problems. Before we commit ourselves to an approach, we should pause to consider alternative approaches. After deciding on an approach, we periodically should attempt to detach ourselves from the details of the problem to reestablish the validity of the overall strategy. If our strategy seems questionable, rather than throw good money after bad, we should reconsider alternative approaches.

12.7 Unrealistic and premature expectations

Users are becoming more and more sophisticated about the capabilities of computers, but still we find many nontechnical, relatively naive users who have unrealistic expectations about computerization. They often do not realize the time, energy, and information that they themselves will have to provide in order to help produce a successful system. They somehow have gotten the feeling that data processing professionals are geniuses who telepathically can work out solutions to ill-defined problems. Furthermore, they expect analysts to tell them what a solution's cost and components will be before the problem, its causes, effects, and overall organizational impacts have been defined. (For those of you who doubt that some users view the computer as magical or mystical, consider your own ability, or inability, to explain your job to relatives or to nontechnical friends.)

The user, however, is not the only important party with unrealistic expectations. The systems analyst typically expects the user to have a detailed understanding of how his business functions. The analyst expects the user to be able to describe clearly the inputs, outputs, and data flows that reflect the specific business application, as well as the logical rules by which decisions are made. We expect users to have a detailed understanding not only of *their* problems and objectives, but also of our tools, including data flow diagrams, flowcharts, and hierarchical system's models. We expect that they can relate to the data processing terms that abound in our specifications and that they can evaluate the cohesion, coupling, flexibility, and maintainability in our systems designs.

But systems development life just is not that easy. Users should not expect magic from systems analysts (garbage in, garbage out); nor should systems analysts expect analytical thinking, detailed understanding of the business, or awareness of the problems in computerization from the users.

The fact that the analyst is aware of data processing's potential and that the user may not be frequently results in a particular problem. When an analyst asks a user, for example, "What do you want the system to do?" he should not be surprised to hear the following reply:

> "How do I know exactly what I want until I see what I get? Once I start using the system, I'll know enough to tell you its weaknesses and strengths. Until then, all of this is just a vague description. I simply cannot envision the possibilities of what you are proposing to me."

Put yourself in the position of the user in the above example, who cannot envision the system merely by reading a description: You might read thousands of descriptions in a newspaper's real-estate section but not know what house you want to buy until you at least see the actual property that seems close to your specifications. If a picture is worth a thousand words, a logical system's model may be worth a hundred discussions with a user.

By creating a logical model of the system, we can eliminate some of the magic and more properly give part of the responsibility for developing a system to the user. The user thus will be more directly involved throughout the entire project. Be aware, however, that logical modeling sometimes creates a problem. Since logical designs are presented to the users long before the actual physical system is available, the users may want the finished product that much faster. They reason, "If it takes you only a month to show us an extensive logical design, why do we have to wait a year for the actual system?"

Top-down implementation, as effective as it is, also has contributed to this problem. Once users see initial versions of the system and they realize how good the ultimate system will be, they tend to demand premature delivery of the entire system. The analyst must take the responsibility to educate users in order to minimize such expectations.

12.8 Personnel assignment

Before the modular design concept was accepted generally, the smallest recognizable predefined body of logic was a program. In those days, designers developed rambling narrative program specifications that involved many interconnected functions. Project managers, unable to extract functional pieces from the overall program specifications, normally assigned one programmer to code an entire program. Experienced programmers were given critical programs to write, and inexperienced programmers were given simpler, less important programs. Each of these programmers typically went to a separate corner and did the assigned task. Experienced programmers asked questions, got clarifications, utilized sophisticated techniques, and produced commendable programs. Inexperienced programmers coded the specifications as they read them, line by line, without question. Inefficient programming techniques were utilized, specifications were misinterpreted (or misstated), and the programs reflected the inexperience of their creators.

More important than the differences in style or in techniques utilized by programmers was the fact that their programs frequently did not fit together. Program-to-program incompatibilities were not discovered until late in the project, when there was no time to meet the deadline. Even worse, a serious problem resulted whenever any programmer quit, because only that programmer was familiar with the program he was writing.

Top-down modular design, coding, and testing provide project managers with more and better options regarding the assignment of personnel on a project. Large programs are broken into a known number of independent modular functions, each with a predictably manageable module size. As a result, programming requirements and assignments can be made on a module, rather than a program, basis. A module test plan can be developed to establish the order in which modules are to be coded and tested. Furthermore, because the functional requirements of every module are known, the creation of test data and the acceptance of the system (or versions of the system) can be coordinated easily with the programming effort.

As already mentioned, the top-down modular approach to systems development provides project managers with the opportunity to assign programming by modular function. Assigning small, unrelated pieces of a problem to a programmer results in good control over the implementation effort. After all, how far can the design drift if small, independent functions are coded and tested one by one? How indispensable can a programmer be if he is coding a large number of manageably small modules, as opposed to a small number of unmanageably large programs? The disadvantages are that if programmers do not frequently and effectively communicate with other programmers on a project, they cannot see the entire picture, or contribute to solving problems outside the domain of their particular assignments, and they do not necessarily learn from each other.

Such problems can be overcome by utilizing a team concept during implementation. Much has been written about the chief programmer team concept. By definition, chief programmers' duties typically include writing functional specifications, designing the system, coding critical modules, supervising and coordinating testing, generating systems documentation, communicating with users, and leading the overall technical team effort. But this approach generally is not recommended for a variety of reasons:[2,3,4]

- Chief programmers cost too much.

- Their salaries and positions do not fit in with the established structure of most organizations.

- Chief programmers normally do not already work for the organization.

- They normally cannot be found, even outside the organization.

- *Egoless* chief programmers do not exist (Have you ever met one?).

We should not, however, discard the team approach because of the difficulties associated with chief programmers. Rather, we must move from the thought that "reading someone else's program is like reading their personal mail,"* so that we can incorporate individual efforts into an overall team strategy. If inexperienced programmers are to learn from their more experienced associates, we must create an environment in which all programmers are encouraged to discuss and critique each others' work freely. If we are not to be continually threatened by the indispensable programmer, we must eliminate him by exposing his work early and often to peer review. If we are to develop maintainable systems, we should involve maintenance programmers in the project as it is being developed. And, if we want to get

*E. Yourdon, "Making the Move to Structured Programming," *Datamation*, Vol. 21, No. 6 (June 1975), p. 54.

programmers involved and enthused about a project, we should give them the broad picture, so that they can contribute to the overall effort, and share in the rewards when the project is a success. There may be some casualties along the way as egotistic, stubborn personalities clash, but the spreading of information and expertise and the overlapping of backgrounds and skills that are by-products of the team approach are benefits that far outweigh the associated risks.

12.9 Software package selection

Software packages are computer programs or sets of computer programs designed to meet common objectives for a variety of users. In general, they either support specific business applications (accounts receivable, general ledger, inventory control, for example) or systems-oriented functions (simulation, compilation, job accounting).[5,6] Figure 12.2 identifies many of the types of available software packages.

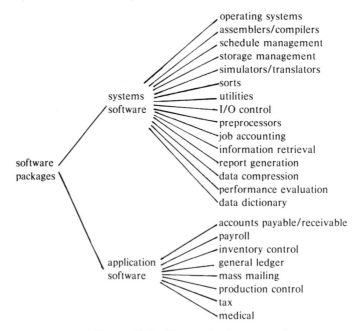

Figure 12.2. Types of software packages.

Software packages are important because they provide useful functions in an economical way. As businesses become more dependent on computers, organizations may not have the technical expertise, time, or budget to develop the software for the applications they want. Nor do they need to, for software packages already exist; they minimize costs because many organizations share in the developmental and maintenance costs of a package. The buyer of a package (following thorough package evaluation) also can be assured that the software product will do the required job and will be flexible, well supported, and well documented.

Costs to develop the software are a major part of the overall costs of a data processing installation. Studies indicate that total software costs equal an estimated one percent of the gross national product and more than 75 percent of total data processing costs. Some of these costs are unavoidable, but large savings can be achieved in many cases by purchasing generalized software packages, frequently for one-fifth to one-tenth of their development costs. [7]

Because of this apparent opportunity for significant savings, software package evaluation and selection have assumed larger roles in automated systems development. Unfortunately, these tasks have presented major problems to systems developers, who frequently are ignorant of the types of packages available. Moreover, so many software packages are on the market that it is difficult to choose the best one to solve the problem. As a result, software packages that would lower overall systems development costs may not be selected; and those that are selected sometimes sit unused on the shelf, despite the considerable time and money invested in their procurement.

One reason for these problems is that systems architects do not treat the evaluation and selection of software packages with the same degree of precision as they do other development tasks. They often rush into package acquisition or in-house software development without systematic consideration of the problem or relative costs of potential solutions. After functional objectives and the alternative approaches to problem solution are identified, systems architects should determine whether software packages to suit their needs are available.

How to distinguish the relative merits and limitations of the many potentially relevant packages is another problem. Many organizations have developed questionnaires to gather information from vendors;[8] a list of typical questions follows:

- Does the basic function of the package meet your needs?

- Will the package run on your computer system?

- What are the software requirements?

- How well do the detailed capabilities of the package match your requirements?

- Is the performance adequate for your needs?

- How flexible is the package?

- In what form is the package delivered?

- How difficult will it be to install the package?

- Will the package be easy to use?

- Is adequate documentation provided?

- What support will the vendor provide?

- What is the operational status of the package?

- What is the total cost of acquiring and using the software package?

- What financial arrangements are offered?

These generalized questions frequently leave as many questions unanswered as they answer because they are not sufficiently detailed. As a result, even after supposedly systematic evaluation, organizations end up with packages with severe incompatibilities with their environment. These questions should be supported with more precise informational requirements, which should be as detailed as the characteristics and limitations of the environment in which the package will be installed.

While there is a danger in establishing detailed lists of questions without knowledge of the specific characteristics of a particular environment, the following list of questions represents the approximate level of detail required to assure that a software package will be suitable. Once these questions have been answered, sample tests can be made, a contract negotiated, and the package finally installed.

What is the background of the vendor?

- What is the main business of the vendor?
- How long has the vendor been in this business?
- How large is the vendor's staff?
- Is it a private or public company?
- What is the prior experience or contractual relationship with the vendor, if any?

What is the background of the software package?

- How long has it been available?
- What are its main features and functions?
- What are its main limitations and constraints?
- Are there different versions of the package?
- Which version is being proposed?
- How often are new releases developed?
- When is the next release planned?
- How long is each release supported?
- Are new releases mandatory?
- Who is currently using the version of the proposed software package?

What are the technical characteristics of the package?

- On what computers and models will it run?
- What programs comprise the package?
- Is the software package modularized?
- Are the modules functionally independent?
- Are there any growth or space limitations?
- Does the package require any other packages?
- Under what operating system version does the package run?
- With what configurations is the package not compatible?
- Is any special equipment or any special feature required or recommended?
- Does the package allow for special user exits?
- In what language is the package written?
- Which compiler versions are required?
- Which linkage editor and options are required?
- What are the inputs and outputs? file descriptions? record descriptions? access methods?
- How does it edit, handle errors, and compute?

What kind of support will be provided?

- What is the lead time for delivery?
- Is there a trial period before acceptance?
- Who provides the test data?
- Who aids in system's conversion?
- Who is the key contact for the account?

- How many support people will be assigned?

- How available and efficient are the assigned support people? percentage of time available? remote or on-site? average response time? average fix time?

- What documentation is provided?

- Can we maintain the documentation by ourselves? Is that desirable?

How are problems identified and ideas exchanged?

- Is there a user newsletter?

- Are there user group meetings?

- Are there diagnostic aids?

- Is there a formal trouble-reporting procedure?

- Who is the vendor key contact?

What are the operational requirements?

- How is the system started and shut down?

- What recovery procedures exist?

- What are the input and output requirements? number of data sets? physical device requirements?' approximate size of data sets? special forms (printing)? density requirements (tape)?

- What is the maximum amount of core required?

- Are there special performance considerations?

- What is the anticipated run-time?

- Are there console operator messages? If so, what are they?

What are the associated costs?

- Initial installation?

- Additional installations?

- Enhancements and modifications?

- Maintenance?

- Training?

- Manuals?

- Any other functions? What are they?

- Choice of lease or buy?

- Choice of terms and conditions?

Review Exercise

1. Identify the types of problems we might expect when a systems development environment is not guided by any systems development methodology.

2. Identify the strengths and weaknesses of a systems development methodology with which you are familiar.

3. How can we help to assure that the users will be able to relate to and evaluate the documents we send to them?

4. How does the spiral approach to systems development differ from the straight-line approach? What are the pros and cons of each approach?

5. How would you distinguish a logical statement from a physical statement? Who cares and why?

6. Why should systems analysts attempt to order the importance of objectives?

7. What is wrong with the systems documentation that you have seen? How would you improve it?

8. Explain the value of face-to-face presentations as opposed to distributing written documents for review and approval.

9. Specify a situation in which you have recognized resistance to change. Why did it occur? Was there any value in that resistance? Could it and should it have been overcome?

10. Identify typical false and premature expectations that both users and systems analysts have had. How do these unrealistic expectations affect systems development efforts?

11. How can systems analysts cope with users who say, "How do I know exactly what I want until I see what I get?"

12. How does the modular design concept affect assignment of personnel in the implementation phase of a project?

13. What are the pros and cons of a team approach to systems development?

14.   How does a team concept relate to the problem of the indispensable programmer?

15.   What are software packages and why are they concerns of a systems analyst?

References

1.   P.R. Lawrence, "How to Deal with Resistance to Change," *Harvard Business Review,* January-February 1969, pp. 4-12, 166-176.

2.   F.P. Brooks, Jr., *The Mythical Man-Month* (New York: John Wiley & Sons, 1975).

3.   F.T. Baker, "Chief Programmer Team Management of Production Programming," *IBM Systems Journal,* January 1972.

4.   P. Naur, B. Randell, and J.N. Buxton, eds., *Software Engineering, Concepts and Techniques, Proceedings of the NATO Conference* (New York: Petrocelli/Charter, 1976).

5.   S. Wooldridge, *Software Selection* (Springfield, Va: Petrocelli/Charter, 1973).

6.   R.A. Smith, "Guidelines for Software," *Journal of Systems Management,* April 1972.

7.   F. Miller, "Selection of Software," *Journal of Systems Management,* July 1977, pp. 12-15.

8.   "How to Buy Software Packages," *Datapro,* Datapro Research Corp., October 1973.

PART 4

Structured Methodology

13 Structured Methodology

The magnitude, complexity, and nature of the job should relate to the level of detail, time, and effort expended in document creation and evaluation.

13.1 Disciplines of structured methodology

Structured methodology is a collection of procedures that provides guidelines for systems development and maintenance. It is based on the structured disciplines of *structured analysis, structured design,* and *top-down implementation with structured programming.* While any one of these disciplines theoretically alone could be applied to systems development, a basic tenet of structured methodology is that these disciplines overlap and are mutually supportive. (See Table 13.1.) Before proceeding with a discussion of structured methodology, let's review relevant aspects of the separate disciplines.

Structured analysis, a disciplined approach to the analysis process (that is, general problem-solving, not just the analysis phase), is applicable throughout the entire systems life cycle. Awareness of classical analytic pitfalls and guidelines to develop precise problem definitions, statements of objectives, estimates, and logic specifications are as applicable in the post-evaluation of systems as they are in determining their feasibility.

Structured design starts with a problem statement, which has been transformed by a data flow diagram into a logical model of the proposed system. Structured design involves decomposing the DFD to the appropriate level of detail, and transforming this logical model into a detailed physical design, including logic specifications to support hierarchical modular structures.

Table 13.1

Relationship of Structured Disciplines to Phases

PHASE/DOCUMENT	DISCIPLINES		
	SA	SD	TDI
1. Request Evaluation	√		
2. Systems Survey/Feasibility	√		
3. Objectives & Priorities Def.	√	√	
4. Physical Alternatives Def.	√	√	
5. Detailed Logical Design	√	√	
6. Top-down Implementation Plan	√	√	√
7. Detailed Physical Design	√	√	
8. Top-down Implementation	√		√
9. Post-evaluation	√		

Top-down implementation is a strategy that usually begins after the modular structure of the design has been completed, but before detailed logic specifications for each of the modules have been created or finalized. Thus, top-down implementation often can start before a structured design has been completed.

13.2 Characteristics of structured methodology

13.2.1 Graphic documentation

Structured methodology recognizes the need for graphic documentation to support systems development and maintenance. For example, in response to management's needs to monitor the schedule, budget, and overall developmental progress of projects, structured methodology recommends regular status reporting, supported by general project milestone charts, cost-analysis diagrams, and detailed activity-time diagrams. To assure that all users have been identified and are involved throughout the developmental process, analysts and project coordinators can use organizational charts, supported by a definition of each department's responsibilities and functions. To minimize potential misunderstandings with program and modular specifications, analysts and designers should supplement complex specification narratives with structured English, IPO diagrams, decision tree structures, or decision tables. All of these graphic tools used in combination with data flow diagrams, data structure diagrams, structure charts, and a data dictionary provide the documentation required for project control, project development, maintenance, system's enhancements, and training of new project personnel. The relationship between many of these graphic tools and the specific phases of structured methodology is shown on the following page in Table 13.2.

13.2.2 Logical and physical design aspects

Structured methodology recognizes that the design process should be separated into two steps: logical design and physical design. Analysts and designers should determine *what* functions the proposed system is to accomplish (logical design) before they identify *how* the system physically will accomplish these functions (physical design). This breakdown has the advantage of providing users with solutions to which they can relate. While users frequently are unable to evaluate highly technical, physical design documents, they can evaluate logical designs expressing functional requirements.

Table 13.2
Relationship of Graphic Tools to Phases

PHASE/DOCUMENT	Proj. Management Charts	Data Flow Diagrams	Data Structure Diagrams	Logic Tools	Structure Charts
1. Request Evaluation					
2. Systems Survey/Feasibility	O	R	O	O	O
3. Objectives & Priorities Def.	O	R	O	O	
4. Physical Alternatives Def.	R				
5. Detailed Logical System	R	R	O	O	
6. Top-down Implementation Plan	R				R
7. Detailed Physical Design	R	O	O	R	R
8. Top-down Implementation	R			R	
9. Post-evaluation	O	O	O		O

O = Optional
R = Required

13.2.3 Versioned implementation effort

Structured methodology recognizes that project deadlines are more manageable and that severe communication problems can be minimized by showing the users a preliminary version of the system as soon as possible. Because many users cannot envision the impact of a system until they start using it, lengthy system's implementation efforts should be broken down into project versions.

ve version provides the users with additional
y. Before the final version is accepted, users
system's versions that process some real input
and p   real output. As a result, the users can begin
to work with and evaluate a subset of the system long before the
final version has been completed.

Breaking a large implementation effort into versions also
provides a reliable measure of project progress. Variances to
schedule estimates are quickly identified, schedule overrun
surprises are minimized, and timely notification to all relevant
parties can be made when expectations do not match what is actually produced.

13.2.4 Structured meetings

Structured methodology recognizes the importance of structured walkthroughs, which are formal meetings in which design
and program specifications, coding, or any systems development
documents are exposed to peer review to identify errors, omissions, and ambiguities. This strategy provides continuous monitoring of a product and results in early refinements and correction, when changes are relatively easy and inexpensive to make.
Other objectives of the walkthrough strategy are to identify
weaknesses in style; to assure modularity, efficiency, and maintainability; and to provide training of inexperienced personnel.

In order to make the most efficient use of time, these
meetings should be structured to be optimally productive. For
example, long, rambling meetings in which attendees get bogged
down in irrelevant details can be eliminated by establishing precise statements of the meeting's objectives and time constraints,
and by involving only a manageable number of participants (usually three to six people). Materials to be reviewed should be distributed in advance of the meeting, and detailed minutes should
be distributed after the meeting to assure that the results are understood by all participants. Any outstanding action from one
meeting should become part of the agenda of a subsequent meeting. In addition, meeting time can be optimized if attendees
concentrate on the identification of problems rather than on the
correction of problems.

13.2.5 Team approach

Structured methodology recognizes the importance of a team approach to systems development. While not specifically advocating the chief programmer team concept defined in the previous chapter, this strategy proposes that the best techniques not only surface but survive as well when people work together. Trainees and junior personnel tend to learn effective tools and strategies when their work is reviewed by more experienced members of the team. They also learn by being exposed to how the more experienced team members approach and solve their problems. A team approach also improves overall systems support, because no programmer or analyst becomes indispensable if others on the team understand what he or she is doing. Perhaps most important, a team approach provides the optimum match between responsibility and capability. Rather than have one or two people responsible for all aspects of systems development, people can be assigned to tasks according to their functional capabilities. For example, Job Control Language experts can help with the JCL; project librarians can support the documentation effort; user groups can help with test data creation; and, as a result, designers can spend more time designing, programmers more time programming.

13.2.6 Need for flexibility

Structured methodology recognizes that guidelines for systems development and maintenance must be flexible to be applicable to the wide variety of systems being developed. Some systems take years to implement; others can be completed in a month. Some systems involve interactions with many users and many departments; others do not. Some systems have flexible deadlines; others must be completed by a specific date. The magnitude, complexity, and nature of the job should relate to the level of detail, time, and effort expended in document creation and evaluation. For example, in systems development, the actual documentation acceptance precedes systems acceptance. But in a maintenance environment, documentation of fixing a bug usually occurs *after* the problem has been corrected and the system has been accepted.

Flexibility also is important in the document review process. When an organization is spending millions of dollars to develop a system, each document associated with each phase of development of the project should be reviewed carefully by all relevant parties. This evaluation may take weeks for each document produced. But if a governing body dictates that your organization make a system do something within two months, your organization will not have the luxury of weeks in which to evaluate a document. Thus, the methodology chosen by the organization should be flexible enough to provide specific guidelines for such an occurrence.

Table 13.3
Phase/Document Applicability Chart

PHASE/DOCUMENT	PROJECT DEVELOPMENT			MODs		Main-tenance
	Small	Medium	Large	Minor	Major	
1. Request Evaluation (Operation Trouble Rpt)	R	R	R	R	R	R
2. Systems Survey/Feasibility		O	R		R	
3. Objectives & Priorities Def.	O	R	R		O	
4. Physical Alternatives Def.	O	O	R			
5. Detailed Logical Design	R	R	R	O	R	
6. Top-down Implementation Plan	O	O	R		R	
7. Detailed Physical Design	R	R	R	R	R	O
8. Top-down Implementation	R	R	R	R	R	R
9. Post-evaluation	O	R	R	O	O	

R = Required
O = Optional

Table 13.3 shows a suggested phased approach for the variety of systems development and support efforts normally undertaken. It indicates a streamlined approach for small project development, minor modifications, and maintenance. Additional documents and checkpoints are required for medium-size projects and major modification efforts. All phases and supporting documents would be required for large project developments.

The difficulty in using such a chart is inversely related to the precision of the definition of the terms used. What is a *small* project? What is the difference between a *major* and a *minor* modification effort? These terms should be defined so that development costs, benefits, impacts, and risks all are considered in light of the relative priorities of an organization. For example, perhaps more attention should be paid to a $50,000 project that has high visibility and must be completed within four months than to a non-critical, $70,000 project with low visibility. From this example, we can see that terms such as small, medium, large, minor, and major relate more to the system's overall impact on an organization than to development costs.

13.3 Control within structured methodology

As suggested by the disciplines of structured methodology, a systems development effort involves the phases of analysis, design, and implementation. However, separating a project into these three phases with document review only at the end of each phase will not provide management with enough control over the systems development process. When millions of dollars are being invested in automated solutions to complex problems, management must establish more regular checkpoints to assure that the overall objectives of the organization are being met.

Structured methodology assures better control because it specifies a phased approach to the systems development process. Each phase identifies detailed functional activities and results in a written document which is reviewed by all designated signatories associated with the effort. Each written document (except for Post-evaluation) becomes a checkpoint at which time designated signatories decide whether to continue or to stop the process. While a phased approach implies a specific, orderly sequence of documents, there are considerable iteration and overlap in the activities represented by written documents. While somewhat distasteful from an auditing or project control point of view, some iteration and overlap are both natural and desirable. A few examples to support this point follow:

- An estimate by definition is a tentative assessment or approximation of the time, cost, or value of a process. Rough estimates are developed early in the systems development process, and continually are refined as specific approaches, activities, and responsibilities are identified.

- The definition of objectives and priorities involves educating the user community with regard to the relative costs associated with response time, turnaround time, equipment, and the trade-offs between time and space. This may involve overlapping the definition of objectives with some tentative physical design to establish a dollar framework for the logical system's objectives.

- Once a general approach to problem solution has been accepted, simultaneous development of the top-down implementation plan and the detailed physical design is desirable. These should be parallel rather than sequential efforts.

- The physical design and the implementation of systems are not necessarily totally sequential efforts. While implementation normally involves the completion of an overall physical design strategy (a hierarchical modular structure for all modules in the system), implementation can be initiated before low-level details of the physical design have been finalized.

Structured methodology is oriented toward a general user, data processing environment. The approach recognizes general departmental entities, their responsibilities, and interdependencies. While no two organizations have exactly the same political and structural environment, every organization involved in complex systems development has some or all of the following environmental entities:

- an *initiating entity,* which starts the process by requesting that a study be undertaken to address an existing or anticipated problem

- a *developmental entity,* responsible for producing an automated or manual solution to a problem

- a *user entity,* comprised of the various users for whom the system is being created

- an *operating entity,* which runs the system once it is automated (sometimes considered to be a part of the user entity)

- a *coordinating entity,* responsible for coordinating documents' distribution, and perhaps assuring that organization standards and policies are met; this entity also may report on the status of all projects within the organization

- an *executive entity,* which resolves conflicts and makes strategic decisions on behalf of the organization

Figure 13.1, on the facing page, shows how the documents of structured methodology flow among the environmental entities of an organization. Once a request for developmental work has been initiated and properly screened, a developmental entity prepares a document (Systems Survey/Feasibility), which is sent to a coordinating entity. The developmental entity and coordinating entity determine which parts of the document are of specific interest to each of the designated document reviewers.

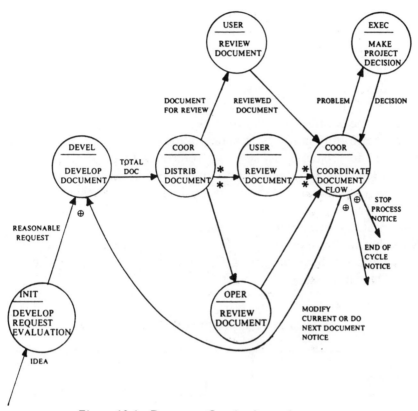

Figure 13.1. Document flow in the environment.

The designated personnel review the document, note approval or disapproval, and return the document to the coordinating entity. If the designated reviewers unanimously have approved the document, the coordinating entity notifies the developmental entity to continue the project by producing the next document. If the designated reviewers unanimously have rejected the document, either the document is sent back to the developmental entity for modification, or the project is stopped. If the designated reviewers have neither unanimously approved nor rejected the document, the executive entity is brought into the decision-making process to resolve the conflict. Once the executive entity has made its decision, the coordinating entity notifies the appropriate parties that the document has been approved or is to be modified, or that the project is to be stopped. This process continues for all documents produced, until the implementation of the system has been approved and the system properly installed.

Proper control of a systems development effort requires clear definitions of responsibility and of authority regarding the effort and its review. Because each developmental effort is unique, the number of required phases and the definition of the user community and its responsibilities should be established early in the project and used as a reference throughout the project. Using phase applicability guidelines established in Table 13.3, we can develop definitions of departmental responsibilities for each appropriate phase. For example, Table 13.4 shows hypothetical departmental responsibilities for a medium-size project. Note that in this example, all phases and documents are required except for the Systems Survey/Feasibility and Physical Alternatives Definition steps.

Table 13.4
Departmental Responsibilities for Sample Project

PHASE/DOCUMENT	User 1	User 2	Syst Devel	Oper	Standards Review
1. Request Evaluation	D				R,A
3. Objectives and Priorities Def.	C,R,A	C,R,A	D	R	R,A
5. Detailed Logical Design	C,R,A	C,R,A	D	R	R,A
6. Top-down Implementation Plan	C,R,A	C,R,A	D	R	R,A
7. Detailed Physical Design	C,R,A	C,R,A	D	C,R,A	R,A
8. Top-down Implementation	C,R,A	C,R,A	D	C,R,A	R,A
9. Post-evaluation	C,R	C,R	D	R	R

A = Approve; C = Consult; D = Develop; R = Review

13.4 Phases of structured methodology

13.4.1 Phase 1: Request Evaluation

The objective of the Request Evaluation phase is to identify an existing or anticipated problem. The output of the Request Evaluation is a preliminary problem definition. The document should be screened internally by the management of the initiating department before being sent to the coordinating entity for distribution and evaluation. Guidelines and responsibilities are listed below:

1.1   Define the problem.

- historical perspective
- problem statement

1.2   Define the problem environment.

- departments
- departmental association with the problem
- key personnel contacts

1.3   Identify any external sources familiar with the problem and/or potential solutions.

1.4   Quantify the effects of the problem.

- lost revenue
- avoidable costs
- operational inefficiencies
- intangibles

1.5   Identify any constraints associated with the problem.

- policies
- procedures
- standards
- laws

- regulations
- deadlines

1.6 Identify any preliminary ideas on potential solutions to the problem.

1.7 Include any additional commentary.

13.4.2 Phase 2: Systems Survey/Feasibility

The objective of the Systems Survey/Feasibility phase is to identify an existing or anticipated problem in an environment and the feasibility of solving the problem. This phase should determine what currently exists, what is wrong with the current situation, who is affected, and the impact of the problem on the overall organization. The document should conclude with an identification of broad alternative approaches to problem solutions and a feasibility statement pertaining to analytic and developmental efforts for potential subsequent work to solve the problem. The output of the Systems Survey/Feasibility should be distributed to designated reviewers for their criticism and comments. The steps of this phase are outlined below:

2.1 Describe the problem, distinguishing causes from effects.
- who
- what
- when
- where
- why
- to what extent

2.2 Identify the current system's user community and all other relevant organizational entities.
- organizational chart
- definition of department charter associated with the problem
- key contacts

2.3 Describe the current system.

- narrative
- physical I/O diagram
- data flow diagrams
- data structure diagrams
- structure chart
- program description
- techniques and procedures

2.4 Summarize the current system's limitations and constraints.

- time
- space
- functions
- error rates (inputs, outputs, and supporting procedures)
- volume
- past business changes
- anticipated changes
- historical maintenance and enhancement costs
- policies
- procedures
- standards
- lifetime costs to date

2.5 Estimate the manpower, supplies, service costs, and schedule needed to complete the next two development phases (Objectives and Priorities Definition and Physical Alternatives Definition).

2.6 Identify preliminary approaches to problem solution.

- for each general physical alternative, identify approximate costs and time for development; extent in meeting objectives; estimation of benefits; potential impact; estimates of maintenance and enhancement costs for the expected lifetime of an alternative

2.7 Develop a feasibility summary.

- indicate whether the situation warrants further developmental analysis

- establish the rationale for the conclusions

2.8 State any outstanding problems in the environment and the action to be taken.

13.4.3 Phase 3: Objectives and Priorities Definition

The objective of this phase is to identify the logical objectives and priorities of the multiplicity of users in an environment. In this phase, systems analysts should be transferring their attention from the current to the proposed environment. The output document will include a preliminary logical design of the proposed system, which establishes priorities within objectives. It should not be interpreted as a proposal for the physical solution to the problem. The output of this phase should be distributed to designated reviewers for their commentary, verification, and concurrence. The following list provides phase guidelines:

3.1 Define objectives of the proposed system that address limitations of the current system.

- precise, measurable statements
- rationale identification
- costs quantification (developmental and lifetime)
- benefits quantification
- objectives/requirements differentiation
- rank and priorities

3.2 Develop levels of data flow diagrams to depict the most extensive logical requirements of the proposed system.

3.3 Develop a data structure diagram to depict data interrelationships required to service inquiries that demand immediate response.

3.4 Identify and rank the types of inquiries to which the system must respond immediately.

3.5 Identify the types of inquiries to which the system will not respond, either immediately or at all.

3.6 Identify any outstanding problems and the corresponding action to be taken.

3.7 Create a preliminary data dictionary.

13.4.4 Phase 4: Physical Alternatives Definition

The objective of this phase is to identify and describe possible alternative physical solutions to the problem. The document produced is essentially a menu of physical alternatives and includes preliminary cost and schedule estimates for each alternative proposed. The document concludes with a comparative evaluation of each alternative to identify which one best meets the objectives of the users. The document should be distributed to designated reviewers, who should regard it as a preliminary physical system's proposal. Designated reviewers should be encouraged to test, rather than accept, the findings in the comparative evaluation subsection of the document. Guidelines follow:

4.1 Describe in general terms the physical system's alternatives to meet the logical objectives of the users, for example.
- 100% manual
- 50% manual, 50% automated
- batch
- on-line day, batch overnight
- on-line, centralized data base

4.2 Tentatively design each physical system's alternative to identify the following.
- key features
- impact on the user community

- handling of changes
- ability to meet users' objectives
- limitations and exposure
- time and cost estimates (developmental and life-time)
- benefit estimates

4.3 Compare and evaluate each physical system's alternative to identify which one best meets the objectives of the users.

13.4.5 Phase 5: Detailed Logical Design

The objective of the Detailed Logical Design phase is to identify the logical processing objectives and requirements, and the logical data structure interrelationships for the proposed physical solution to the problem. In a sense, the output of this phase is a result of superimposing the definition of objectives and priorities of the users (Phase 3) over the limitations associated with the choice for physical solution (Phase 4). The output of this phase also can be thought of as the functional specification in that it is not physical, is specific, and will form a basis for determining whether the system has met the users' objectives during the Post-evaluation phase. The document produced should be distributed to all relevant parties for their review and concurrence, and should be based on the following guidelines:

5.1 Summarize the current system and its limitations (attach or refer to background information).

5.2 Develop precise, measurable statements to describe the agreed-upon objectives of the proposed system.

5.3 Describe the proposed system.
- general capabilities
- organizational responsibilities
- impact on the organization

- expected improvements
- limitations and exposure
- time and cost estimates (developmental and life-time)
- benefit estimates

5.4 Develop levels of DFDs for the proposed system.

5.5 Develop a DSD for the proposed system.

5.6 Identify the various inquiries associated with the proposed system.
- system input (logical)
- system response
- volume estimate (peaks and valleys)
- response time
- security and privacy requirements
- relative importance

5.7 Refine the data dictionary to reflect terms used in data flow and data structure diagrams.
- data flows
- data elements
- processes
- attributes
- entities
- aliases
- abbreviations
- cross-reference capability

13.4.6 Phase 6: Top-down Implementation Plan

The objective of the Top-down Implementation Plan phase, which is outlined below, is to provide order and continuity for all activities required for physical design and implementation. If the project is a major new development or major enhancement effort, it should be broken down within the plan into a number of sub-projects or versions, each with gradually increasing functional capabilities. The plan should specify personnel by activity, and include refined cost and schedule estimates for the entire project and its sub-projects. Activities and their associated schedules and cost estimates identified in this phase will be viewed as part of the basis for evaluation during the Post-evaluation phase. The versioned implementation plan should be developed with the users of the proposed system. The Top-down Implementation Plan should be distributed to all designated reviewers for their evaluation and concurrence.

6.1 Describe the total plan.

6.2 Identify the specific activities.
- physical design
- programming
- manual procedures
- test data creation
- test acceptance
- training
- document preparation
- standards enforcement
- project management
- walkthroughs
- reporting
- conflict resolution

6.3 Identify specific personnel for each activity listed.

6.4 Estimate the manpower, time, and costs involved in each activity.

6.5 Estimate the time and costs for the total plan.

6.6 Devise a series of implementation versions with gradually increasing functional capability.
- activities for each version
- responsibilities for each version
- cost and schedule estimates by activity
- functions and capabilities to be delivered

- hardware, software, and human interfaces to be tested
- testing and acceptance methods
- effect of the version on the users
- cost and schedule estimates for entire version

13.4.7 Phase 7: Detailed Physical Design

The objective of the Detailed Physical Design is to develop the physical structure of both the application system's design and its associated data base design if required. The document of this phase also will reflect a refinement of physical design objectives, which will be part of the basis for evaluation during the Post-evaluation phase. The output of this phase should be sufficiently precise to provide all of the information required to implement the physical design. This information should include structure charts and program and module specifications for coding the system, as well as methods and procedural specifications for all supporting manual functions. The output also should include a data dictionary, identifying and cross-referencing all terminology used in system's documentation. The output of the Detailed Physical Design phase should be distributed to designated reviewers for their review, criticism, and concurrence. Following is an outline for this phase in the systems development process:

7.1   Refine design objectives.
- functional capability
- volume capability
- resource requirements
- response time
- execution time
- error rates
- maintenance costs

7.2   Define the modular program or system structure.

- a nested hierarchy of modules
- one-entry, one-exit modules
- functionally independent modules
- manageably small modules
- simplified interfaces
- limited data communication
- graphic depiction of module logic specifications expressed by structured English, decision tables, or decision tree structures

7.3 Develop the physical data base design.
- data structure diagram
- security and privacy provisions
- resource requirements
- resource allocation
- supporting procedures
- anticipated inquiries lists
- response-time requirements

7.4 Physically package the system.
- programs
- job steps
- overlay structure
- file and record définitions
- file access requirements
- resource allocation

7.5 Identify specifications for methods and procedures at the boundaries between the automated system and its supporting manual functions.
- console operator manual
- user manuals
- data preparation procedures
- output data interpretation guidelines
- test data creation
- test data acceptance

- standards enforcement
- systems release procedures
- training requirements

7.6 Complete the data dictionary.
- glossary of terms
- definition of data flows, data elements, logical processes, programs, modular processes, subroutines, aliases, and abbreviations
- cross-reference capability

13.4.8 Phase 8: Top-down Implementation

The objective of the Top-down Implementation phase is to meet the functional, operational, cost, and schedule objectives of the users. The effort should follow the guidelines established in the Top-down Implementation Plan (Phase 6), as well as the guidelines presented here. The system should be coded following structured programming techniques. System's coding and testing should proceed essentially in a top-down fashion, and should utilize continually expanding test data. Major interfaces should be tested early in the process. Module stubs should be used to facilitate testing, rather than driver or skeleton programs. Testing should be done incrementally, by adding a manageably small number of untested modules to a working subset of the system, rather than by the traditional phased approach to testing (unit, integration, subsystem, and system). If the project has been broken down into sub-projects or versions, each should be coded, tested, and accepted one by one. The output of this phase is a series of listings, test cases, and testing results by version. Supporting documentation for each version, such as procedure and console manuals, also is produced. These documents should be distributed to all designated system's reviewers for their review and concurrence. Guidelines for this phase follow:

8.1 Identify the scope of the implementation being distributed for review and concurrence.

8.2   Identify the testing criteria and results.

8.3   Identify the personnel associated with implementation acceptance and their specific responsibilities.

8.4   Produce all documentation to support the scope of the implementation being accepted.
- program listings
- user manuals
- console operator manuals
- data preparation procedures
- data dictionary additions and revisions

8.5   Identify any existing cost and schedule overruns; for any overruns, develop a specific plan for subsequent development.

13.4.9 Phase 9: Post-evaluation

The objective of Post-evaluation is to determine the extent to which the objectives of the system have been met. A second aspect of this phase is to identify the nature, scope, and impact of implementing modifications that have been postponed until after the system has been put into operation. The output should include an identification of the best and worst techniques and strategies utilized during the project, and suggestions for improving the systems development approach. The document should be distributed to designated reviewers. No commentary or concurrence is required. Guidelines for this phase follow:

9.1   Identify the objectives of the system as stated in prior documents.
- Detailed Logical Design
- Top-down Implementation Plan
- Detailed Physical Design

9.2   Identify all variances to the stated objectives.
- nature

- extent
- impact
- reason for occurrence
- cost

9.3 Evaluate any outstanding requests for system's enhancements.

- impact on logical functions (DFD)
- impact on physical design (structure and I/O charts)
- impact on the data base (DSD)
- impact on user community
- impact on performance
- impact on overall system's maintainability

9.4 Identify any recommendations for upgrading the system's functions or performance.

9.5 Develop a plan to meet recommendations, unmet objectives, and postponed modifications.

- functions
- activities
- functions and capabilities to be delivered
- cost and schedule estimates
- estimate of benefits
- versions (optional)

9.6 Summarize the systems development effort.

- best techniques
- worst techniques
- avoidable costs
- lost opportunities
- suggestions for improvement

Review Exercise

1. What is a systems development methodology?

2. What are the major differences between traditional systems development methodologies and structured methodology?

3. Describe the departmental relationships in your organization as they pertain to systems development efforts. How do they differ from organizational relationships defined in structured methodology?

4. How do the disciplines of structured analysis, structured design, and top-down implementation with structured programming overlap? How can they be separated?

5. What is a phased approach to systems development?

6. What is a versioned approach to systems implementation? How can it help to measure progress?

7. What is meant by the terms *logical* and *physical* design?

8. Identify the major characteristics of the last meeting that you attended. How could that meeting have been improved? What were its best aspects?

9. Define an existing problem, using the Request Evaluation guidelines. Evaluate that definition. What additional information could you have defined? How could you improve the definition of the problem?

10. Describe the applicability of the data flow diagram to the phased approach of structured methodology.

11. How does the data dictionary relate to the phased approach of structured methodology?

APPENDIX

Systems' Review

- Systems' Review receives all documents from Administration Control.

- It reviews and comments on documents, noting approval or disapproval for all documents except the PER.

- All PERs are checked against corporate goals. PERs normally are relatable to the corporation's goals; if they are not, they are sent via Administration Control to the Executive Review Council. For all Problem Evaluation Requests, Systems' Review assigns a department for Problem Definition responsibility.

- Systems' Review returns all documents to Administration Control.

Executive Review Council

The Council meets once per month. It is made up of a designated representative from each of the departments that comprise the organization. Among its other duties, the Council has two decisions to make regarding the systems development process, as follows:

1. When PERs are unrelated to the goals of the organization as established by the Organization Yearly Plan, the Council determines whether the Plan should be modified to support the PER.

2. When a document produced as part of the systems development life cycle does not receive unanimous approval, the Council assigns a status indicator to the document: *approved, to be modified,* or *rejected* (the last indicator meaning that the project must be stopped).

Assigned Group

Systems' Review assigns to a particular group responsibility for the development of systems development documentation and implementation. The assigned group is notified by Administration Control of its responsibilities regarding the creation of documents. The assigned group completes a document and returns it to Administration Control. Documents having an invalid format as well as those requiring modification will be returned to the assigned group by the Administration Control Department.

You now have the necessary information to begin drawing your data flow diagram for the Porterhouse Shrimp Corporation's flow of systems documentation. Go to it!

Glossary

access time
: the time required for a device to receive or transmit data after receiving the associated command.

address
: a label, name, or number identifying a location or device where information is stored.

alias
: a label, name, acronym, or symbol that is synonomous with a more commonly accepted name.

analysis
: the process of examining, identifying, or separating a complex to isolate and define the relationship of its components.

analysis phase
: the front-end phase or phases of the systems development life cycle prior to physical design.

analysis process
: the process of identifying and examining problems associated with the entire systems development effort.

assemble
: the process of converting a nonmachine language program into instructions in a machine at specified locations.

attribute
: a property or characteristic of an entity; a data element in a data base organization.

attribute pointer
: an attribute in one entity that is used as a pointer to a particular set of information (a record) in another entity.

black box
: a system, program, module, or component with known inputs, known outputs, and a generally understood function, but with unknown or irrelevant detailed contents.

bottom-up testing	a testing strategy in which bottom-level modules in the modular hierarchy are unit-tested first and then integrated into higher-level modules; usually contrasted with top-down testing.
bubble chart	an alias (synonym) for a data flow diagram; see data flow diagram.
business analyst	see business systems analyst.
business systems analyst	a person, usually from the business or applications area, who represents the user's interests, by specifying business objectives, priorities, and constraints, and who actively collaborates in the systems acceptance effort.
central transform	an aspect of transform analysis; central transforms are central systems' functions that transform highly processed input data into logical output data.
code	a system of symbols and rules to represent information, such as data and the instructions to process data.
cohesion	a measure of intramodular strength; the degree of functional relatedness of processing elements (instructions) within a module.
coincidental cohesion	the lowest and least desirable level of cohesion; refers to a module whose processing elements have no constructive relationship.
communication	the process of transferring information among people, devices, or locations.
communicational cohesion	an intermediate level of cohesion; refers to a module whose processing elements all operate on the same input data and/or produce the same output data.
conservative top-down implementation	the process of coding and testing a modular hierarchy after the entire modular design has been completed; usually contrasted with radical top-down implementation.

control module	the main module; the module at the top of a modular hierarchy; the module that coordinates and manages the activities of subordinate modules; sometimes thought of as a driver, executive, monitor, supervisor, or coordinate module.
conversational mode	the process of communication in which terminal entries elicit responses from a computer; the terminal is inhibited while responses are being prepared.
coupling	a measure of intermodular strength; the degree of dependence between modules.
data base	a file of interrelated data that are stored together to serve one or more applications and that are independent of programs using the data.
data base administrator	a person responsible for assuring the integrity of a data base.
data base designer	a person responsible for developing the physical data base design.
data dictionary	a repository of data about data and processes associated with a particular systems development effort; includes a glossary of terms, data characteristics, process descriptions, and cross-references.
data element	a unit of data defined in a data dictionary; a field on a record.
data flow diagram	a graphic tool that represents data flow and transforms in a process. Also known as DFD, data flow graph (DFG), or bubble chart.
data structure diagram	a graphic tool used to represent entities, attributes, and data interrelationships in a data base that are required to retrieve data immediately in response to inquiries. Also known as DSD.

debug	to detect and to correct errors in a procedure, system, program, or module, or in a piece of equipment.
decision table	a graphic tool used to describe the conditions and actions in a problem in a tabular format.
decision tree structure	a graphic tool used to describe the conditions and actions in a problem in a tree-structure format.
design	to plan the form and method of a solution; see logical design and physical design.
disk storage	a storage device with magnetic recording on a flat, rotating disk.
driver	a primitive simulation of a control module or program used in bottom-up testing (unit-testing).
dummy module	see stub.
enhancement	see modification.
entity	a general category about which information is recorded in a data base environment. For example, customer information is in the customer entity.
entity identifier	see search argument.
execute	to carry out an instruction or to perform a routine or set of routines.
executive module	a high-level module in a modular hierarchy, which manages and coordinates the activities of lower-level modules.
factoring	the process of decomposing a program or system into a modular hierarchy.
fan-out	see span of control.
feasibility study	a front-end study in a systems development project that identifies the nature and scope of a problem to determine the likelihood of a

system being built within established constraints to solve that problem.

filtering   a special type of data transformation (in a data
flow diagram) in which the data are not physi-
cally changed but our knowledge about that
data is changed. A filtering transform, for example, might separate raw data into two
streams: an in-sequence stream and an out-
of-sequence stream.

flexibility   a measure of the degree to which a program,
system, or module can be modified or used in
a variety of ways.

flowchart   a graphic tool to show the sequence and control of program or module logic.

functional cohesion   the highest and most desirable level of cohesion; refers to a module whose processing ele-
ments all contribute to only one function.

functional specification   a document detailing a system's functions to
be performed, the objectives and require-
ments to be met, the performance criteria by
which the system will be judged, and the constraints within which the system must be
delivered; normally, forms the basis of an
agreement about specifications between systems development personnel and users, and is
the basis for the physical design effort.

functionality   a synonym for cohesion.

heuristic   a guideline, which gives general direction in
problem-solving, but which is not guaranteed
to yield the best results.

HIPO   an acronym for *H*ierarchy-*I*nput-*P*rocess-
*O*utput; an IBM documentation technique for
modular systems.

immediate response   a response to an inquiry within a time period
usually measured in seconds.

implementation plan

a plan that identifies all activities, responsibilities, objectives, and cost and time estimates to implement a system or program.

incremental testing

a testing (implementation) strategy in which a manageably small module or group of modules is added to a working subset of the system and tested together; the process continues until the entire modular structure has been assembled and tested (usually contrasted with phased implementation or bottom-up implementation).

inquiry

a request for information from storage that often requires an immediate response from an automated system.

interface

a common boundary between two devices, subsystems, programs, or modules.

interface complexity

a factor that influences coupling between modules: the greater the complexity, the higher the coupling (an undesirable trait).

ey attribute

see search argument.

logical cohesion

one of the lower, and, consequently, less desirable levels of cohesion; refers to a general-purpose module whose processing elements are associated by similarity in function.

logical design

one phase in a phased approach to systems development; the process of transforming users' desires, objectives, and requirements into ordered, measurable specifications, supported by graphic tools that show logical inputs, outputs, and processes.

logical pointer

a pointer (graphically depicted by an arrow) to identify the relationship between entities in a data structure diagram. It indicates the ability to gain access to information in one entity by defining a key attribute in another entity.

logical record

a set of information limited, not by format or size, but by the nature of the information.

logical statement   a statement that can be implemented physically in more than one way.

loop   a construct of structured programming; a coding technique in which instructions are repeated, sometimes with modification of instructions and/or data values.

maintainability   the extent to which a program or system can be corrected easily and cost-justifiably when bugs are discovered during the program's or system's productive lifetime.

maintenance   the phase in a system's life cycle following development, acceptance, and installation.

modifiability   the ability of a system to be changed or enhanced to meet changing objectives and requirements; flexibility, changeability.

modification   a change or enhancement to a system.

module   an identifiable piece, component, or contiguous sequence of instructions; a component of a program.

morphology   shape, particularly with respect to the structure of modular hierarchies and organizational structures.

multi-programming   the process of overlapping or interleaving the execution of several programs.

normal connection   a reference from one module to the name of another module (as opposed to a pathological connection).

off-line operation   an operation that requires some human intervention between data origination and computer processing.

on-line operation   a direct operation of equipment by a computer; requires no human intervention between data origination and computer processing.

packaging	the process of transforming logical functions into modules, programs, job steps, and so on; process of identifying and defining distinct physical units for machine execution.
pathological connection	a reference from one module to a label or data within another module (as opposed to a normal connection); influences coupling.
personnel subsystem	the part of data processing systems composed of people; the inputs, processes, outputs, standards, procedures, methods, guidelines, documentation, manuals, and training requirements to assure that people can interface effectively with a data processing system.
phased implementation (testing)	a form of implementation (testing) in which several partially tested modules are combined at once and tested together for correctness.
physical design	one phase in a phased approach to systems development; the process of transforming a functional specification or logical design into plans for programs, modules, and subroutines, including detailed program, module, and subroutine specifications.
physical record	a storage record with a specified format and of predefined size.
physical statement	a statement that can be implemented physically in essentially only one way.
priority	the order of importance in which requests, entries, and jobs will be handled or processed.
procedural cohesion	an intermediate level of cohesion; refers to a module in which all processing elements are involved in a set of procedures not necessarily connected by any required sequence or continuity of data.
program	a group of related instructions or processing elements that meet the requirements stated in a program specification; also to code and debug the instructions or processing elements of the program specifications.

| program specification | a precise description of a program's input, output, and processing requirements; the directions given a programmer by a designer. |

pseudocode — a tool to specify program, module, or policy logic without conforming to the syntactical rules of any programming language; see structured English.

radical top-down implementation — the process of designing, coding, and testing one level of a modular hierarchy before handling and processing subsequent levels.

record — a group of related data, facts, or fields of information that is treated as a unit; see logical record and physical record.

recursion — continued repetition of operations or of a group of operations; the act of invoking a module as a subordinate of itself; a recursive module is one that calls itself.

reliability — a measure of the quality of a program, system, or equipment.

response time — the elapsed time between the generation of an inquiry at a terminal and the receipt of a response at the terminal; includes transmission time to and from the computer and computer processing and access time.

routine — a sequence of instructions that carry out a function; part of a program or module.

scope of control — a structural parameter of a module consisting of the module and all its subordinates.

scope of effect — for a decision, the collection of all modules containing any processing that is conditional upon that decision.

search argument — the attribute value or values used to identify and retrieve a unique set of data within an entity; an entity identifier or key attribute.

sequential cohesion	a strong and consequently desirable level of cohesion; refers to a module whose processing elements require a specific sequence and continuity of data. A module is sequentially cohesive if the output data of one processing element serve as the input data to the next processing element.
software packages	generalized computer programs that are designed to meet common objectives for a variety of users.
span of control	a module characteristic; the number of modules immediately subordinate to a module; fan-out.
specification	a precise definition of objectives, requirements, and constraints; a technical, engineering description for hardware; a definition of inputs, outputs, and processes.
standards	approved rules and required practices for controlling the technical performance and methods of personnel involved in systems development, modification, and maintenance.
storage	a general term to describe any device capable of retaining information.
structure chart	a documentation technique to show the hierarchy of modules and interrelationships among modules in a program or system; similar to a HIPO table of contents.
structured analysis	a philosophical, top-down approach to all phases of the systems development life cycle, featuring graphic tools, an awareness of classical problems, and a structured methodology.
structured design	guidelines and techniques to determine which modules, interconnected in which way, will best solve a system's problem.

structured English	a type of pseudocode; a tool to supplement or replace narrative specifications; a representation of policies, procedures, and logical rules in a disciplined English language format, conforming with the logical constructs of structured programming.
structured methodology	a collection of procedures based upon structured disciplines, which provides guidelines for a phased approach to systems development and maintenance.
structured programming	guidelines and techniques for writing programs as a nested set of single-entry, single-exit blocks of code, using a restricted number of constructs.
stub	a primitive implementation of a subordinate module (normally used in top-down testing).
switch	a device or program condition that controls the flow of logic in a program or system.
system	interdependent devices, rules, and/or procedures organized to form an integral whole to achieve a common purpose.
systems analysis	the examination, identification, and evaluation of the components and interrelationships involved in systems development efforts; the definition of problems, objectives, requirements, priorities, and constraints of systems, plus identification of cost, benefit, and schedule estimates for potential solutions.
systems analyst	a person who does systems analysis and who acts as the interface between users and designers and programmers.
tape storage	a medium (paper or magnetic) for storing information that can be used as computer input or output.
temporal cohesion	an intermediate although not necessarily desirable level of cohesion; refers to a module whose processing elements are all related in time.

terminal	an input-output device capable of sending or receiving data to or from a system of which it is a part.
testing	a phase in the systems development process following design and implementation; a process of demonstrating that a module, program, or system functions as specified.
top-down design	an informal modular design strategy in which the major functions of a system are identified and expressed in terms of lower-level subfunctions; this process of functional decomposition is repeated until the subfunctions are at a sufficiently low level such that their implementation can be expressed easily.
top-down implementation	a testing (implementation) strategy in which major, high-level modules are tested before more detailed, lower-level modules; usually requires stubs to simulate low-level modules in testing high-level modules.
transaction	a signal, event, or unit of data that triggers or initiates some action or sequence of actions.
transaction analysis	a modular design strategy based on an analysis of the sources, types, and actions of transactions in a system.
transaction center	a portion of a system that identifies transactions and calls the appropriate detailed modules to complete the processing of each transaction.
transaction-centered design	see transaction analysis.
transform analysis	a modular design strategy in which program or systems structure is derived from an identification of the inputs, outputs, and transformations (processes) required to transform inputs into outputs; analysis of data flow and transformations.
transform-centered design	see transform analysis.

user	a person, persons, or organizational entity expected to derive benefits from the development of a system; a client of the systems development effort.
user analyst	see business systems analyst.
versioned implementation	an implementation effort (plan) that breaks down an overall project into sub-projects or versions so that the effort can be judged by "inch-pebbles," not milestones.
white box	an element whose contents must be understood in detail in order to be utilized.

Index